WITHDRAWN

INDUSTRIAL ENERGY MANAGEMENT FOR COST REDUCTION

INDUSTRIAL ENERGY MANAGEMENT FOR COST REDUCTION

by

THOMAS E. SMITH

**President
Sunrise Technology Inc.
Energy Engineering, Planning, Design
Lihu, Hawaii**

ANN ARBOR SCIENCE
PUBLISHERS INC
P.O. BOX 1425 • ANN ARBOR, MICH. 48106

Preface

The days of limitless fossil energy reserves are past. Not only are our present energy reserves exhaustible, but the tapping of new sources has become extremely costly. My aim in this work is to present the basic considerations for an effective Industrial Energy Management Program so that any energy management team can understand clearly the methods for reducing energy usage.

It is not my intention to concentrate on the subject of engineering in the strictest sense, but rather to relate the basic engineering data that are essential to meet the challenge of operating a facility. As a company grows, problems arise involving alternate energy sources, retrofit, and design criteria that should and must be solved.

My message is to the responsible executive who wishes to plan, actively encourage, and use energy-effective techniques. The conservation of energy at the individual user's level presents an opportunity to decrease energy expenditures and reduce operating expenses while efforts are made to preserve our fast-depleting resources.

The time has come when action programs must be instituted to avoid being forced eventually into reaction programs. Escalating energy costs are putting pressure on management budgets. However, enlightened management through sound energy management programs can ease the impact of inevitable budgetary problems.

It is imperative that management today, using all available resources, be prepared to face the problems of energy shortages and rising costs. Managements can act now or wait to be forced into reacting. The choice is theirs.

THOMAS E. SMITH P.E.

Acknowledgments

Over the course of writing this text and executing the illustrations, I have been helped by many people. I am most grateful for this. Those who have given assistance are Florendo Ancheta, Jr.; the American Society of Heating, Refrigeration and Air-Conditioning Engineers Inc. (ASHRAE); Betty J. Bell; Carrier Air-Conditioning Company; G.N. Wilcox Memorial Hospital, Lihue, Kauai, HI; H & C Metal Products; Vari-Cool Division; Nancy Scibetta; Darlene Smith; Sunrise Technology, Inc.; Trane Air-Conditioning Company; the United States Department of Energy; York Air-Conditioning Company; the State of Hawaii; the Department of Energy; and Thomas E. Smith and Associates. Charts courtesy of Dubin/Mindell/Bloome Associates, Consulting Engineers.

 **Thomas E. Smith, P.E.,** is president of Sunrise Technology, Inc., an energy engineering consultation firm in Lihue, Hawaii. He earned his Master's Degree in Public Administration and his Bachelor of Science Degree in Mechanical Engineering from the University of Illinois. He is involved in facility management and consultation, particularly with respect to energy management, in various fields including health care and industry.

He is a member of the State Energy Committee dealing with conservation and alternate energy sources, and of several professional societies including National Society of Professional Engineers, National Association of Energy Engineers, International Society of Solar Energy, American Section of Wind Energy Engineers, American Society of Hospital Engineering, American Society of Mechanical Engineers and the International Solar Energy Society.

To All My Family,
Especially my wife Darlene and son Keith

Table of Contents

Introduction

It is a pleasure to introduce this book for one simple reason. It is loaded with useful specifics instead of broad generalizations about energy management.

The good old days, when energy was dirt cheap, are gone. Those who doubt this need only read the short first chapter of this book which collects the key facts and figures on the future supply of oil, coal, and natural gas. The escalation in the cost of energy shows no sign of abating. Thus more and more the cost of energy becomes a significant element in competition for product markets and in bottom line performance. The prudent manager now recognizes that no longer can one just buy energy, its use must be effectively managed.

This book does not begin or end, either literally or figuratively, with a discussion of general principles of management in the context of the energy crisis. The tenth chapter does make fundamental points about applying management principles in planning, organizing, and controlling energy usage. Chapters two through nine, however, detail specific steps, techniques and methods which can be utilized to reduce expenditures for energy.

Heating and cooling systems are major energy users and often present significant opportunities for energy conservation measures. The third chapter of this book discusses these opportunities, addressing the use and value of flue gas analyzers, when and how to isolate off-line boilers, the considerations involved in replacing existing boilers with modular boilers, and how to reduce blow-down losses.

The fourth chapter is a virtual compendium of maintenance procedures for improving energy efficiency. Pinpointed are the working elements of various kinds of industrial equipment that should be inspected with a view to improving energy efficiency through more informed and better maintenance practices. The fifth, sixth, and ninth chapters deal with water use and pumping equipment and present techniques for improved performance as well as practical ideas for design and operation modifications. Chapter seven explains steps that can be taken to optimize the use of electricity. Refrigeration equipment is an enormous energy consumer. Chapter eight discusses the critical balance between efficient refrigeration and reducing energy costs and how to achieve it. The practical uses today of non-depleting resources, such as solar energy, are discussed in chapter eleven.

Measures to reduce costs through energy conservation sometimes require at the outset significant capital investment. Chapter twelve explains how to use economic analysis tools to reach sound decisions on whether to make such capital investments.

The last chapter of the book presents two management case studies which graphically demonstrate what can be accomplished through energy management.

Effective Energy Management for Plant and Facilities is a book for those who are committed to achieving tangible *results* from an energy management program.

Tom Watson

Mr. Watson, a partner in the law firm of Morgan, Lewis & Bockius in Washington, D.C., is an adviser to utilities and industry on energy problems and planning. He has been actively involved in the national energy legislation in the Congress as well as in proceedings at the Department of Energy.

CHAPTER 1

The Energy Outlook

ENERGY CONSUMPTION VS. AVAILABILITY

The United States consumes about 58 percent of the earth's energy output, but has only 6 percent of the world's population. The demand for energy is at an all time high and is spiralling in the United States and in other countries.

Several forms of energy are being used — principally coal, petroleum, natural gas, natural gas liquids and electricity. Many Americans consider energy to be inexhaustible. Apathy toward energy savings is fostered by the relatively low cost of fuel. Some buildings were designed and constructed with only one thought — initial cost. This has resulted in a myriad of buildings that by today's standards are excessive energy consumers.

The present supply of energy is not unlimited. Estimates range from 15 years for natural gas to 300 years for coal. As sources of energy expand, gaps will develop, brought about by population growth and increased demand. After nonrenewable energy supplies are depleted, the world must turn to more permanent sources. Although there are several, the four most significant today are wind, nuclear, solar, and geothermal.

In 1960 the total United States oil consumption was approximately 9.7 million barrels per day. Of that amount, only 1.9 million barrels per day were imported and 0.2 million barrels per day were exported. Conversion of coal production to million-barrels-per-day of energy shows that 1960 coal production was equal to about 5.3 million barrels of oil and the gas supply to 5.9 million barrels, mostly from domestic sources.

The 1970s saw some startling revelations. The 1960-1970 decade was noted for its massive expansion (by more than 300 percent) of natural gas consumption and the more than doubling of electricity use from the equivalent of 3.4 million barrels of oil per day to 6.9 million. Oil use increased from 9.7 million barrels a day to about 13.9 and coal from the equivalent of 5.3 to 7.4 million barrels of oil per day to 11.8. The first major nuclear and geothermal energy usage was developed in 1970.

A key consideration is the efficiency effects of input vs. output. During the 1960-1970 decade, lost energy vs. used energy was about 2 percent. By 1971, this trend had been reversed, with an increase of 1 percent in used energy.

Influential factors in the years 1973-1975 were the oil embargo and the business recession. Between 1970 and 1975, energy was utilized at a rate less than that prior to 1970. This appears to have been brought about by the oil embargo, recession, higher cost of energy, and conservation efforts. While consumption leveled off, United States' production of oil and natural gas shifted downward primarily because of a decline in natural resources.

Because of existing conditions, energy's future is a completely different matter from its past. Due to heavy energy use, change cannot come rapidly but must occur as reserves dwindle.

Projections for the 1980s are not promising. Oil use is expected to rise 30 percent over 1970, with the major amount supplied by imports, and coal use is expected to increase by 27 percent, but natural gas should increase only slightly (approximately 3 percent) because of its limited availability. Of prime importance will be a 13 percent escalation in natural gas imports. Nuclear power, while soaring 30 times over 1970 production levels, will comprise only about 9 percent of total energy output.

Energy consumption prospects through 1990 indicate that the United States will use 25 percent more fuel than in 1980. Coal demand levels will be up some 30 percent, increased imports will help natural gas usage to decline 15 percent below 1980 estimates, and solar and geothermal power will grow 180 percent as technologies advance. However, in order for geothermal energy usage to be cost effective, breakthroughs in the current — and costly — technology must occur.

With the continuing rise in energy consumption, alternate sources more abundant must be found. Even though stronger conservation measures will be in force in the 1980-1990 decade, a change in usage patterns is required. While alternate sources are needed desperately, development of new sources has been deterred by environmental problems relating to nuclear safety, offshore oil drilling, and strip mining.

With its increased dependence on foreign imports, the United States is much more vulnerable than it was during the 1973 oil embargo. Business has been plagued with high capital cost for conversions, and the cost of energy — oil in particular — has risen drastically.

COAL

The production of all fossil fuels should peak before the year 2000, assuming that available data for world supplies are correct, the projected increase in usage is valid and alternate sources meet 12 percent of total energy demand.

Surveys in 1970 indicated there were 3.2 trillion tons of coal in the ground. Of this amount, approximately 1.6 trillion tons were considered recoverable. About half of that total was determined from actual mapping and exploration to depths of 3,000 feet, with the balance in unmapped or unexplored areas as deep as 6,000 feet. Information on coal reserves for the rest of the world is qualitative rather than quantitative.

Coal consumption in the United States has varied from year to year. However, with the oil embargo of 1973 and the federal government's pressure to convert to coal, consumption appeared to be on a steady rise. The peak forecast for the year 2000 is based on the assumption that by then the majority of electric-power generating plants will be using coal.

The estimate of 1.6 trillion tons of recoverable coal is based on its presence in areas known to have the most favorable characteristics for reserves of sufficient magnitude to support large-scale mining operations.

The National Petroleum Council in 1973 prepared estimates of economically available reserves in the chief coal-producing regions of the United States. Later studies by the NPC indicated that the amounts of recoverable surface reserves in the western states should be increased because of the availability of strip mining sites and shallow coal beds.

Known coal reserves susceptible to mining practices and economics are sufficient to last at least up to the beginning of the next century. Even though coal resources may be abundant, however, substantial advances in mining technology will be influenced by changes in comparative patterns of fuel utilization and demand.

The following list of world coal resources was published by the National Coal Association, Washington, D.C., 1972:

Table 1-1 Location of Worldwide Coal Reserves

Country	% of World Total	
ASIA		
U.S.S.R.	19.9	
People's Republic of China	16.7	
India	1.0	
Japan	0.2	
Others	0.1	
		37.9
NORTH AMERICA		
U.S.A.	48.2	
Canada	1.4	
Mexico	0.1	
		49.7

Table 1-1 continued

Country	% of World Total	
EUROPE		
Germany	4.8	
U.K.	2.8	
Poland	1.3	
Czechoslovakia	0.3	
France	0.2	
Belgium	0.1	
Netherlands	0.1	
Others	0.4	
		10.0
AFRICA		
Republic of South Africa	1.1	
Others	-	
		1.1
AUSTRALIA		
Australia	1.0	
Others	-	
		1.0
SOUTH & CENTRAL AMERICA		
Colombia	0.2	
Venezuela	0.1	
Others	-	
		0.3
		100

NATURAL GAS

The use of natural gas in the United States is on the decline, primarily because of the unavailability of natural resources, and is expected to decrease by 1.2 percent a year through 1985.

During this period, a 2 percent growth is expected in residential consumption — individual residences, small multiple dwellings, and larger residential units with individually metered apartments. For many years, estimates of future natural gas supplies developed by numerous individuals and organizations varied greatly simply because of different objectives and meanings. In June 1966, to achieve a more unified industry-sponsored effort, the Gas Industry Committee, representing the American Gas Association, the American Petroleum Institute, and the Independent Natural Gas Association of America, suggested supporting a study of the future natural gas supply. In October 1966, the committee set up an organizational structure that provided for broad, diversified representation from other organizations. The Gas Committee Report, "Potential Supply of Natural Gas," first issued in November 1973, showed the potential supply of natural gas in the United States as of December 31, 1972.* The committee's report was updated in 1976. The report was based on the premise that natural gas yet to be discovered or to be added in existing fields would be recoverable commercially in the future, assuming adequate but reasonable field prices and normal advances in technology. Potential gas estimates cover quantities in addition to and not duplicative of proved reserves estimated by the American Gas Association.** The supply of natural gas in the United States is estimated at approximately 600 trillion cubic feet. Twenty-five percent of the potential supply is below 15,000 feet and 12 percent lies under water 250 to 2,000 feet deep. Of the total potential supply, 32 percent is in Alaska; however, problems exist involving transportation and distribution.

With the extreme demands that have been placed on natural gas, the expectation of continued growth and high usage appears to be past. While consumption has been increasing, limited resources and lack of technological advances indicate that natural gas as a major source is limited and other, more available sources will have to be explored and utilized.

---

*AGA, Proved Reserves Report 27 (May 1973).

PETROLEUM

Petroleum products accounted for more than 40 percent of total energy demand in the United States through 1970. From 1970 through 1973 this demand increased at a rate of approximately 8 percent per year. The oil embargo of 1973 slowed the consumption increase, but since 1974 usage has increased to slightly more than the standard preembargo 8 percent growth rate.

As of 1946, the total oil demands in the United States were somewhat less than 5 million barrels per day. By 1970 the demand had grown to more than 15 million barrels per day. In 1976 consumption was estimated to be 22 million barrels per day.

Many factors contributed to this increase:

- increased population

- industrial expansion

- increased per capita income

- revised credit policies

- more automobiles

- more appliances

Several petroleum associations have projected a 3 percent growth rate for petroleum products consumption between 1976 and 1980, and slightly greater consumption from 1980 to 1985. As in the past, gasoline was projected to be the principal product. Economic factors such as higher earnings, vacations, early retirements, and greater longevity were expected to offset the effect of the declining population growth rate on consumption of petroleum products.

Diesel fuel use was expected to expand, with the growth rate of highway consumption increasing by 10 percent a year. If alternate energy sources are not put into effective use, this could result in a greater demand for power generated by turbine, especially for peak demand. With the increase in environmental regulations, diminishing natural gas supplies, and an increased demand for electricity, fuel oil requirements have risen sharply, with an estimated increase of 71 percent expected by 1980. The following tables show factors affecting demand for petroleum products:

Table 1-2 Gasoline Demand for Passenger Cars in the U.S.

Factors affecting demand	1946	1955	1965	1970	1975*
Annual passenger car gasoline					
Barrels per person	4.9	7.4	8.6	10.3	12.6
Barrels per household	19.2	17.4	29.3	33.6	39.6
Barrels per passenger car	16.7	16.2	15.3	15.8	17.4
Passenger cars (millions)	28.2	52.0	75.1	89.8	106.3
Person per car	5.1	3.2	2.6	2.3	2.0
Cars per household	0.8	1.1	1.3	1.4	1.7
Gasoline demand (million barrels per day/billion dollars GNP+	9.6	9.2	7.2	6.2	5.2
Population (millions)	141.9	165.9	194.6	204.8	214.8

*Estimated
+GNP = Gross National Product

Source: Report 2, Census Bureau, Series D Projection, U.S. Department of Commerce (Washington, D.C., 1972).

Table 1-3 Economic Assumptions for Projecting Demand for Petroleum Products

Factor	1970	1980	Per annunm % gain 1970/1980
Population (millions)	204.8	226.7*	1.1
Households (millions)	62.9	74.7	1.7
Gross national product (billions of constant 1958 dollars)	722	1,090	4.2
Federal Reserve Board Index of Industrial Production (1957-1958 = 100)	176	273	4.5

*Estimated

Source: Census Bureau, Series D Projection, U.S Department of Commerce.

The time has come when the United States must look in new directions and redefine its priorities. As a nation, it not only must look into new technology, but also must increase conservation efforts. With this alteration in energy usage patterns, the nation can put its total system into proper balance and maintain its energy integrity for centuries to come.

The true value to society is not the gross energy available, but the net energy, which is what remains after the costs of obtaining and concentrating that energy are subtracted. Calculations of energy reserves intended to project years of supply are measured in terms of gross energy rather than net energy available and may in fact be of shorter duration than often projected.

Societies compete for economic survival. According to the principle expressed by E.O. Trotsky at the United Nations Conference on Energy (1974), "Systems win and dominate that manage their useful total power from all sources and flexibly distribute the power toward needs affecting survival." During times when energy flows have been tapped and no new sources exist, Trotsky's principle requires that systems that win not attempt fruitless growth, but instead use all available energies in high diversity energy growth. The chance and opportunities are ours. Will the nation accept the challenge that faces it?

Chapter 2

Holding the Line
on Energy Costs

CONSERVATION PLANNING

Critical shortages of fossil fuel supplies compel industrial organizations to create energy conservation programs that will reduce consumption and generate cost savings. Complete and comprehensive plans should be outlined, with the lowest cost conservation measures first on the list, followed by modifications requiring greater investment.

Before an organization can establish sound programs, it must study sources of energy waste. Some fuels may be used directly to heat or cool buildings; however, most fuels are converted into steam that heats or cools, depending upon climatic conditions.

Electricity is purchased chiefly for lighting and driving motors, fans, and other types of machinery. On a dollar basis, approximately 65 percent of electricity is used for heating, ventilating, and air-conditioning, and for operating various machines, 25 percent is used for lighting, and the remaining 10 percent is used for miscellaneous activities.

Since the first concern is the implementation of low cost measures, the energy conservation program should begin by examining the electrical rate structure.

Control over retail rates of electric utilities is in the hands of state Public Utility Commissions (PUCs). The states established PUCs with the logic that a utility is a legally authorized monopoly and as such must be subject to some degree of public control. The control usually is limited and primarily covers revenue.

When the utility decides it requires more revenue to enable it to meet the needs for expanded facilities, it develops a case in favor of a rate increase and asks the PUC for rate changes or increases that would be beneficial to the system and its management. Upon receipt of this request, the commission may order an independent audit of the applicant, which usually is followed by public hearings. Utility revenues then are monitored to assure that the rate changes are justified.

Some rate structure changes or increases can have a dramatic effect upon operating budgets. Plant administrators should be completely familiar with rights and protection granted by the law. A copy of the state's PUC statutes should be obtained and read.

Any questions regarding the laws or regulations should be addressed directly to the commission. Once familiar with its legal rights, a company should not hesitate to exercise them. For example, if a utility is requesting a rate change that would affect a company's operation, the firm should seek answers to the following questions:

1. What would the rate increase mean to this organization?
2. How do the facility projections compare to those of other similar organizations?
3. How do data on past usage patterns compare?

RATE STRUCTURES

Most electric and gas utilities have several rate structures; some have ten or more, such as residential, all-electric residential, general, heating, temporary, supplementary, and outdoor lighting. These are subject to one or more additional provisions, such as power factor, auxiliary and emergency service, primary service, high voltage service, and space heating service.

Utility companies should be contacted for information on their rates, services, schedules, and attached provisions. Most utilities have customer service representatives who will review operations to suggest ways of economizing. However, the individual business must initiate the action. At one plant, managers meet with their electric utility representatives once a month for lunch. Following each meeting, the staff returns with information that is helpful in terms of energy cost savings and planning.

When an organization has determined what utility services are available, a survey of electricity use should be performed for each separate service location. At a minimum, the following should be determined:

1. rate schedules used
2. maximum demand and period of occurrence
3. power factor
4. monthly and annual energy consumption
5. average cost per kwh
6. service voltage level and secondary use level
7. transformer and equipment ownership

It also is advisable to talk with other customers of the utility who are in similar positions to determine their rate schedules, use characteristics, and amounts being paid.

A critical review of at least one current bill for each service should be undertaken, and the following questions should be asked:

1. Is the organization receiving its discounts?
2. Is the bill computed properly?
3. Why is the facility billed according to this rate?
4. Why does the plant not qualify for a better rate?
5. Should metering be consolidated?

A satisfactory answer to each question should be obtained. Local utility representatives may be contacted for assistance in answering these questions and requested to review each service. They should be asked the advantages of each rate. It should not be assumed that the utility's calculations, or even rates shown, are correct. Mistakes often are made, so it is necessary to check figures.

As an example of what such a survey and review can yield, one hospital found that changes in electrical service over a five-year period resulted in three out of seventeen services' not being billed properly or not being billed at the most advantageous rate. The review also demonstrated that the hospital's total service was being metered improperly. An immediate billing correction resulted in enough short-term savings to pay for the cost of the study.

In another case, a department store was being billed on a standard commercial schedule. A review of usage patterns indicated that the store could switch to an industrial rate, since its air conditioning load was more than 50 percent of its total kwh demand. A rate switch was made, producing savings of $39,960 a year.

A watchful and questioning eye on rates can achieve substantial savings. However, a review of rates alone is not all that is to be accomplished. It is necessary to realize that the use of any form of energy involves the waste of energy, regardless of efficiency factors. When energy is converted from one form into another, a loss of BTUs (British Thermal Units) is involved. An efficient operation obviously will cut energy losses.

ENERGY CONVERSION PROCESS

To understand better how energy is used and wasted in the typical energy conversion process, consider the following processes:

- the conversion of electricity or steam into chilled water for cooling
- the conversion of a fuel or steam into electricity
- the conversion of fuel into steam or hot water

In a steam system, energy losses occur typically in the approximate percentages shown in Figure 2-1.

Figure 2-1 Energy Losses in Steam Systems

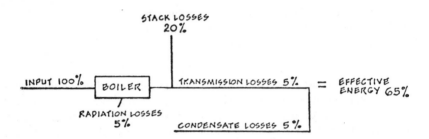

In evaluating each system and/or subsystem, an accurate analysis must be made to arrive at the net energy available after losses. This calculation requires expertise that in some cases may be available within a facility. In other cases, it may be necessary to contact similar organizations for their views or seek the services of a consultant. In any event, it is necessary to obtain the best possible guidance.

SURVEY OF PHYSICAL PLANT

To further the concept of energy management and evaluate areas for potential savings it is best before conducting a survey to assemble basic facts of the physical plant. Enterprises have changes in personnel, and it should not be assumed that all employees are familiar with the entire facility. Take the view that no one person knows where everything is and how various components of the system function.
Below are preliminary steps found useful in preparing for the survey.

1. Designate a physical location as an Energy Management Center. It may be in an existing part of the plant offices or in a newly designated space. The center could be a Plan Room that already exists. It should not be less than 10 feet × 14 feet with a 6-foot table, a desk, and filing facilities.
2. Obtain several copies of a comprehensive site plan; mount one on the wall. Assemble copies of drawings of mechanical and electrical systems and basic architectural outlines. Assemble copies of automatic temperature control diagrams. Assemble shop drawings of major energy-consuming equipment and prepare an index of where they can be found. Prepare thumbnail sketches of buildings and sections. Indicate general types of windows and installed shading. Indicate whether shading is operable. Typical entry: ''Roll shades; always up.''

3. Identify and locate all meters, submeters, and areas served. Locate all boilers, chiller equipment, hot water generators, kitchens, laundries, and incinerators. Obtain copies of electrical demand records from the utility. If district steam is used, obtain similar steam records.

4. Prepare a preliminary list of air handling (AH) systems, types of systems, areas served, method of control, and size designation (Cfm, design MBH heating, tons cooling). Typical entries:

   a. AH-1 variable air volume, thermostat each room, 30,000 Cfm, inlet vane damper control from static pressure regulator, economy cycle.

   b. Operating room system: all outside air, preheat and reheat controls, automatic steam valves.

   c. Accounting offices: exhaust fan for toilet areas, 5,000 Cfm, no central controls.

5. Make a preliminary ranking of known systems according to their energy requirements. List big users first. Prepare a preliminary list of main heat generating systems. If boilers are utilized, indicate type of fuel, type of boiler, boiler rating, method of control. Typical entry: "Three steam generators (Type "D" package units), #4 fuel oil, 40,000 pound-per-hour rating, air-atomized dual-fuel burner (gas/oil), positioning controls, equipped with steam flow recorders." Prepare similar information for water chillers for air conditioning systems, kitchens, laundries, and incinerators.

6. Ask operators what they feel are the constraints to saving energy. Typical entries: "Dampers on economy cycle don't work." "Insufficient relief air." Investigate methods of communication and operation for systems that can be shut down on some appropriate basis. Example: Who operates room air conditioning controls? Talk to personnel directly affected by the energy intensive systems and get their ideas (and feelings) on how to save energy.

ENERGY SAVING TIPS (ESTs)

The ESTs are "resources" that can help managers understand and locate energy-saving opportunities. They should be used to perform a comprehensive facility survey and to investigate ways employees can save energy. The ESTs are for use mainly in existing buildings and systems. They also identify areas that require a high level of energy efficiency in new construction.

Suggestions for actions that will result in energy savings are listed below:

ESTs Listing

- system shutdown during unoccupied hours
- addition of return air
- automatic temperature controls

- economizer cycle
- preheater operation and control
- modification of terminal reheat systems
- use of enthalpy control
- room air conditioners
- energy efficiency tests for chillers
- energy-saving opportunities for chilled water plants
- steam waste prevention
- control of radiator systems
- piping insulation
- domestic hot water
- laundries
- motor shutdown
- demand control/centralized monitoring and computer systems
- lighting
- power factor improvement
- elevators
- window and door leakage
- window modifications
- shading control for windows

System Shutdown During Unoccupied Hours

Savings will accrue in the following ways when shutting down air-handling systems:

1. Not bringing in outside air for ventilation will avoid the cost of heating and cooling outside air.
2. Stopping the fans will save the cost of electricity for operating fan motors.
3. Lowering the temperature during winter periods by combining a night cycle with the shutdown will save additional energy.

With thorough analysis, shutdowns can be made without negative effects on comfort and on maintenance costs.

Many air handling systems appear to have been designed originally with the assumption that almost every area would be a sensitive one. This assumption includes the following rules for equipment design and operation and has a direct bearing on energy usage:

- Ventilation air is required round-the-clock.
- High efficiency air filtration is required.
- One hundred percent ventilation air is required.
- Every system must run continuously.

Some areas of a plant require compliance with these rules. On the other hand, many areas more closely resemble an office building, a school or a warehouse in function. Systems serving such areas often can be shut down. However, care must be taken so that only nonsensitive areas will be affected. The following are a few guidelines that must be considered for shutdown candidates:

- All areas served by the air handling systems must be unoccupied for the shutdown period.
- Exhaust systems for hazardous chemicals and materials must continue to operate (some makeup air may be needed at all times).
- Existing pressure differentials (air paths) in the buildings must not be altered materially.

An air-handling unit for an administration area used as an outpatient clinic serves as an example. The space is unoccupied at night (and probably on weekends). The system is operated when no one will be affected by temperature changes resulting from shutdown. Night air-conditioning is not required in summer and space temperatures can be reduced during winter nights.

A thorough knowledge of the system, including details of operating schedules and procedures, existing controls, and space use is required to determine the best method for the night shutdown control. Control may be accomplished through manual operation, time clock, central monitoring system, or computer. The savings in energy consumption often pay for additional controls in a few weeks. Specific methods and extent of implementation will vary. In some instances, fans must be operated continuously. As a minimum, eliminate outside air during unoccupied periods. If the air-handling system is the sole heat source, intermittent fan operation can provide the minimum heat needed (controlled by a separate space thermostat set at reduced temperature).

When establishing a shutdown with reduced-temperature night cycle, consider modifying the temperature controls so that the air-handling system will start up in the morning with reduced outside ventilating air for an hour or two. This will prevent simultaneous peak use of heating or cooling energy to bring the space back to design temperature and to heat or cool outside air. Such a start-up cycle is critical where a demand charge of purchased energy, such as steam, is involved. The concept of night shutdown (actually any shutdown during unoccupied hours) can be extended to areas occupied only a few hours a week and served by independent systems, such as an auditorium. Essential to the success of this type of change is development of communications between users and operators so that schedules can be coordinated with actual needs. When considering night shutdown, the operation of all systems in the same building must be understood. Physical plant complexity may indicate the need of expert assistance.

Chapter 3

Heating and Cooling

INDUSTRIAL HEATING

The typical heating system operates between 50 and 60 percent of its full load for more hours per year than at higher or lower loads, so the optimum operating point of heating boilers should be selected to give the highest efficiency spread over this range.

If the installation is composed of three 8×10^6 BTU/hr. boilers and the maximum heating load is reduced to 18×10^6 BTU/hr., then the boilers should be adjusted for maximum efficiency at 70 percent of the rated output. This gives three possible operating points at maximum efficiency:

1. one boiler at 8×10^6 BTU/hr. $\times 0.7 = 5.6 \times 10^6$ BTU/hr. or 31 percent full load
2. two boilers at 16×10^6 BTU/hr. $\times 0.7 = 11.2 \times 10^6$ BTU/hr. or 62 percent full load
3. three boilers at 24×10^6 BTU/hr. $\times 0.7 = 16.8 \times 10^6$ BTU/hr. or 93 percent full load.

Indicators of maximum combustion efficiency are stack temperature, percentage carbon dioxide (CO_2), and percentage oxygen (O_2).

Primary and secondary air should be allowed to enter the combustion chamber only in regulated quantities and at the correct place. Defective gaskets, cracked brickwork, broken casings, etc., will allow uncontrolled and varying quantities of air to enter the boiler and will prevent accurate fuel/air ratio adjustment. If spurious stack temperature and/or oxygen content readings are obtained, inspect the boiler for air leaks and repair all defects before final adjustment of the fuel/air ratio.

When substantial reductions in heating load have been achieved, the firing rate of the boiler may be excessive and should be reduced. Consult the firing equipment manufacturer for specific recommendations. Reduced firing rate in gas and oil burners may require additional bricking to reduce the size and shape of the combustion chamber.

The cost of fuel/air ratio adjustments and firing equipment modifications will vary with each application, and quotations should be obtained from qualified contractors.

Analyze Flue Gas

The efficient combustion of fuel in a boiler requires an optimum fuel/air ratio providing for a percentage of total air sufficient to ensure complete combustion of the fuel without overdiluting the mixture and thereby lowering boiler burner efficiency.

Optimum combustion efficiency varies continuously with changing loads and stack draft and can be approached closely only through analysis of flue gases. Information required to allow continuous adjustment of fuel/air ratios is: (a) flue gas temperature and (b) flue gas CO_2 or O_2 content.

Indicators are available that measure CO_2 and stack temperature continuously and give a direct reading of boiler efficiency. These carbon dioxide indicators provide boiler operators with the requisite information for manual adjustment of fuel/air ratio and are suitable for smaller installations or operations where funds for capital improvements are limited.

A more accurate measure of efficiency is obtained by analysis of oxygen content than by measurement of other gases such as carbon dioxide and carbon monoxide.

The cross-checking of O_2 and CO_2 concentrations is useful in judging burner performance more precisely. Due to the increasingly widespread need for multi-fuel boilers, however, O_2 analysis is the single most useful criterion for all fuels since the O_2 total air ratio varies within narrow limits.

For larger boiler plants, consider the installation of an automatic continuous oxygen analyzer with trim output that will adjust the fuel/air ratio to meet changing stack draft and load conditions.

Most boilers can be modified to accept automatic fuel/air mixture control by flue gas analyzer, but a gas analyzer manufacturer should be consulted for each installation to be sure that all other boiler controls are compatible with the analyzers.

It is important to note that some environmental protection laws might place a higher priority on visible stack emissions than on efficient optimization of fuel combustion, especially where fuel oil is concerned. The effect of percent total air on smoke density might prove to be an overriding consideration and limit the approach to minimum excess air. All applicable codes and environmental statutes should be checked for compliance (See Figure 3-1).

Figure 3-1 Effect of Flue Gas Composition and Temperature on Boiler Efficiency

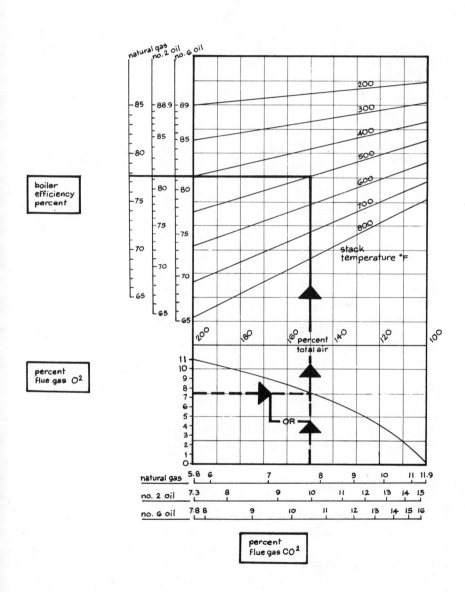

The cost of adding flue gas analysis instrumentation and controls will vary with each case but for order of magnitude the following tabulated costs may be used.

- O_2 analyzer w/meter readout $ 2,000 - $ 3,000
- O_2 analyzer w/chart recorder   3,000 -   5,000
- O_2 analyzer w/trim link control   4,000 -   6,000
- Full metering w/O_2 trim link   11,000 - 16,000

Example:

1. A boiler burning No. 2 oil rated at 20,000,000 BTU/hr. has a yearly oil consumption of 425,000 gallons.
2. A CO_2 meter is installed and reads 11 percent CO_2 in the flue gas.
3. Stack temperature is 650° F.
4. Entering the graph at the 11 percent CO_2 point for No. 2 oil and intersecting with a stack temperature of 650° F gives an efficency of 78 percent; air percentage of 138 indicating 38 percent excess air and an O_2 percentage of 6 also can be read.
5. Installation of an O_2 analyzer with automatic fuel/air ratio control will reduce O_2 content to 3.5 percent and stack temperature to 530° F.
6. Using the graph again, the new efficiency is 83 percent.
7. Yearly savings due to increased boiler efficiency

$$= \text{Original yearly consumption} \times \frac{(\text{New Efficiency} - \text{Original Efficiency})}{\text{New Efficiency}}$$

$$= 425,000 \times \frac{(83 - 78)}{83} = 25,600 \text{ gallons/year}$$

at 33¢/gallon = $8,448

8. The installation cost is approximately $16,000 and obviously is economically advantageous without further economic analysis.

Isolate Off-Line Boilers

Light heating loads on a multiple boiler installation often are met by one boiler on-line with the remaining boilers idling on standby. Idling boilers consume energy to meet standby losses that can be aggravated further by a continuous induced flow of air through them into the stack and up the chimney.

Unless a boiler is scheduled for imminent use to meet an expected increase in load, it should be secured and isolated from the heating system by closing valves and from the stack and chimney by closing dampers.

Figure 3-2 Boiler Bypass Valves and Regulating Orifice

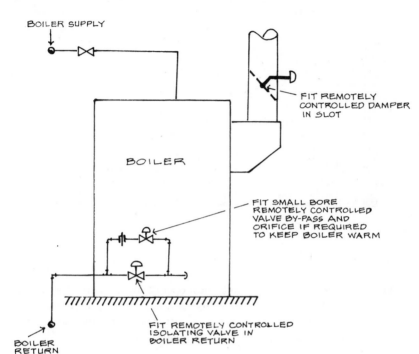

Large boilers can be fitted with bypass valves and a regulating orifice to allow a minimum flow through to keep them warm and avoid thermal stress when brought on-line again. (See Figure 3-2.)

If a boiler waterside is isolated it is important that air flow through the stack be prevented because backflow of cold air can freeze the boiler.

The cost of isolating off-line boilers will vary from case to case and quotations should be obtained from local contractors.

Replace Existing Boilers with Modular Boilers

Heating boilers usually are designed and selected to operate at maximum efficiency when running at their rated output, and have lower efficiencies at reduced output. (See Figure 3-3.)

Typical heat load distribution over a heating season shows that full boiler capacity is required for only small periods and that for 90 percent of the time the heat load is 60 percent or less than full load.

Figure 3-3 Typical Heat Load Distribution for 250-Day Season in 6,000/ Degree-Day Zone

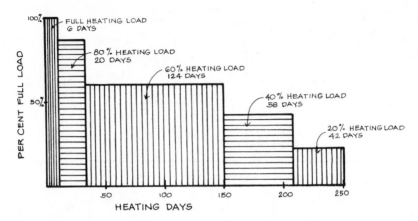

Large capacity boilers in single units must operate intermittently for the major part of the heating season. Although high-low firing capabilities may reduce cycling, the boilers can reach their design efficiency for only short periods, resulting in low seasonal efficiencies. (See Figure 3-4.)

Figure 3-4 Effect on Cycling to Meet Part Load

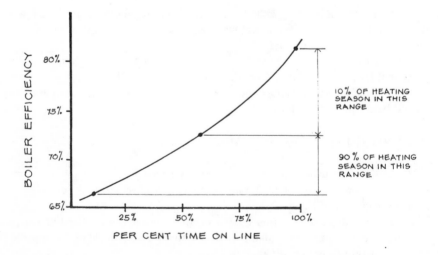

A modular boiler system composed of multiple boiler units, each of small capacity, will increase seasonal efficiency. Each module would be fired at 100 percent of its capacity only when required and fluctuations of load would be met by firing fewer boilers longer.

Each small capacity unit has low thermal inertia (giving rapid response and low heat-up and cool-down losses) and will be running at maximum efficiency or will be turned off.

Boiler seasonal efficiency may be improved from 68 percent to 75 percent in a typical installation where single unit large capacity boilers are replaced by modular ones. This represents a saving of approximately 9 percent of yearly fuel consumption.

Where the boiler plant has deteriorated to the point where it is at or near the end of its useful life, consider replacement with modular boilers sized to meet the reduced heating load.

Modular boilers are particularly effective in buildings that have intermittent short occupancy such as churches. They provide rapid warm-up for occupied periods and low standby losses during extended unoccupied periods.

Preheat Combustion Air to Increase Boiler Efficiency

Preheating primary and secondary air will reduce the cooling effect when it enters the combustion chamber and will increase the efficiency of the boiler. It also will promote more intimate mixing of fuel and air that will add to boiler efficiency.

Figure 3-5 Efficiency Increase with Preheated Air

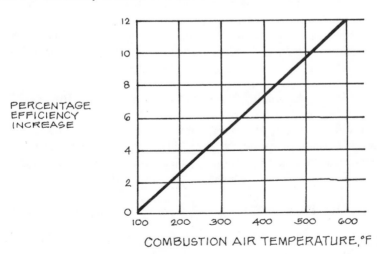

In most boiler rooms, air is heated incidentally by hot boiler and pipe surfaces and rises to collect below the ceiling. This air can be used directly as preheated combustion air by ducting it down to the firing level and directing it into the primary and secondary air inlets.

Waste heat reclaimed from boiler stacks, and blowdown or condensate hot wells, also can be used to preheat combustion air. (See Figure 3-5.)

Boiler efficiency will increase by approximately 2 percent for each 100° F increase in combustion air temperature.

Combustion air can be preheated up to 600° F for pulverized fuels and up to 350° F for stoker fired coal, oil and gas. The upper temperature limit is determined by the construction and materials of the firing equipment, and manufacturers' recommendations should be obtained.

Heat exchange from flue gases to combustion air may be made directly using static tubular or plate exchangers or rotary exchangers. Heat exchange also may be made indirectly through runaround coils in the stack and combustion air duct.

Waste heat from flue gases, blowdown, condensate, hot wells, etc., or from solar energy may be used to preheat oil either in the storage tanks (low sulfur oil requires continuous heating to prevent wax deposits) or at the burner nozzle. .

Oil must be preheated to at least the following temperatures to obtain complete atomization:

- No. 4 oil - 135° F
- No. 5 oil - 185° F
- No. 6 oil - 210° F

Heating beyond these temperatures will increase efficiency, but care must be taken not to overheat, or vapor locking could cause flame-outs. The increase in efficiency obtained by preheating oil could be as high as 3 percent but depends on the constituents of the oil, and recommendations should be obtained from the oil supplier.

Change Steam Atomizing Burners to Air Atomizing Burners

Air atomizing burners are considerably more efficient than steam atomizing burners and should be considered as an alternative when replacing obsolete or defective steam atomizing burners. In some circumstances, they may be cost effective on their own merit without waiting for scheduled replacement times.

Reduce Blowdown Losses

The purpose of blowing down a boiler is to maintain a low concentration of dissolved and suspended solids in the boiler water and to remove sludge to avoid priming and carryover. There are two principal types of blowdown — intermittent manual and continuous. Blowing down is accomplished by opening the valve at the rear of the boiler, then opening the blowdown valve, which removes the concentration of dissolved and concentrated solids. To help the removal of suspended solids, the surface blowdown at the water line also should be opened after the rear blowdown has been performed. Manual or sludge blowdown is necessary for the operation of the boiler regardless of whether continuous blowdown is installed. The frequency of manual blowdown will depend on the amount of solids in the boiler makeup water and the type of water treatment used. Continuous blowdown is a steady energy drain, as the makeup water must be heated. Energy loss from blowdown can be minimized by monitoring with automatic blowdown control and heat recovery systems.

Automatic blowdown controls monitor the conductivity and PH of the boiler water periodically and blow down the boiler only when required to maintain acceptable water quality. Further savings can be realized if the blowdown water is piped through a heat exchanger or flash tank with a heat exchanger.

In a 100,000 lbs/hr., 600 psi boiler with a maximum boiler water concentration of 2,500 ppm total dissolved solids, the blowoff will be 8 percent of the makeup water of 3,500 lbs/hr. The total heat in the blowoff is 1,660 MBH (thousands of BTUs per hour). A system using heat exchanger only and with a 20° F terminal difference will recover 90 percent total heat in the blowoff, or 1,494 MBH. Adding a flash tank operating at 5 psig with a heat exchanger having 20° F terminal difference will recover 93 percent of the total heat in the blowoff, or 1,544 MBH. The percent of heat recovery will change with boiler operating conditions. The best recovery range is 78 percent for a 15 psig boiler to 98 percent for a 300 psig boiler.

As demonstrated previously, there are many ways to control effectively the energy consumption in heating systems. Outlined below are basic guidelines to reduce energy usage in these systems.

Guidelines to Reduce Energy Used for Heating

1. Lower Indoor Temperature and Relative Humidity Levels in Heating Season

    a. Install a seven-day dual thermostat to operate the oil, gas, stoker, or electric heating elements. The thermostat should be set to maintain the temperature levels during unoccupied periods and to reduce levels when the building is unoccupied at nights, on weekends, and on holidays.

    b. Install a seven-day thermostat to control pumps for forced circulation hot water systems, and automatic control valves for hot water or steam systems when boilers are operated by aquastat or pressure-trol.

    c. Install a seven-day timer control to reset operating aquastat or pressure-trol for dual level settings.

d. Install additional thermostats for individual zones where duct or piping systems permit control of individual zones and where necessary at zone control valves. Operate controls on a seven-day cycle.

e. Provide room thermostat and automatic damper controls and dampers in supply air duct systems for additional zoning to permit further reduction of temperature levels in noncritical areas.

f. Where radiator valves or duct dampers are missing, install manually operated ones to control or shut off heat supply to noncritical areas in portions of the building.

g. Install occupied/unoccupied, on-off switch to control water supply to humidifiers to permit shutdown at night.

h. Remove and relocate central humidifier serving the entire building and install duct type or package room humidifier to serve only those zones that require humidification. Shut off humidifiers at night as per (g) above.

i. Do not operate refrigeration systems or introduce outdoor air for cooling in winter to reduce indoor temperature levels to the heating set point of 68° F. If gains exceed losses, allow space temperature to rise to 78° F.

j. Where room thermostats control both heating and cooling systems, exchange them for units with a dead band level.

k. Provide locking devices on all room thermostats to prevent tampering.

l. Where supply duct dampers are adjusted to reduce air flow into the space during the heating season, provide damper to reduce return air from the space.

m. Relocate room thermostats to the most critical area and rebalance air or water system to reduce temperature and humidity levels in other less critical areas.

n. Where window air conditioner units or through-the-wall units provide heating as well as cooling, provide seven-day temperature control thermostats to operate the heating elements.

2. *Reduce Heating Load Due to Ventilation and Makeup Air*

a. Provide seven-day timer to operate automatic fresh air damper control (install controller if missing) or to shut off all outdoor air for ventilation during unoccupied periods where separate outdoor air supply fans exist.

b. Install automatic damper control (if missing) for operation with seven-day timer.

c. Add separate ventilation fan for zones requiring outdoor air and shut off damper to separate unit serving the entire building.

d. Reduce operating time of exhaust systems and, where interlocks with outdoor air makeup systems have been or will be installed, control the exhaust-outdoor air makeup unit or damper with seven-day timers.

e. Utilize waste heat or recovered heat from exhaust systems and other equipment to temper outdoor air supply to reduce heating load.

f. Install charcoal filters or other devices to control supply air quality and reduce the amount of outdoor air for ventilation. Check with code authorities before making the change.

g. Modify exhaust hoods to reduce the ventilation as follows:

- By inspection, determine whether existing exhaust hood is of the simple open type and by measurement determine quantity of air exhausted. If exhaust hood is of the open type, install baffles to allow reduction of exhaust air quantity without reducing the face velocity at edge of hood.
- If baffles are installed, reduce rate of exhaust and by measurement determine the new quantity.
- Install a makeup air system to introduce outdoor air equal in quantity to that exhausted.
- Do not heat makeup air to more than 55° F and introduce as close as possible to the hood in several positions around the perimeter to promote even air flow.
- Consider using reclaimed hood exhaust heat or solar heat as the energy source for makeup air.
- Do not cool or dehumidify makeup air.

COOLING

Heating, ventilating, and air-conditioning have a significant impact on a building's total energy consumption. Each cubic foot of air brought into a building must be either heated or cooled and in some cases humidified and/or dehumidified. Local building codes now require 20 to 25 to 30 percent minimum outside air. This becomes expensive for plant managers.

To be completely effective, codes will have to be changed. The fastest approach appears to be through energy regulation that would be followed by code changes. Under normal conditions, it takes approximately 5 cubic feet of air per minute to supply the required oxygen and other dilution required per person. In evaluating building code requirements, specifying cubic feet per minute per person rather than cubic feet per minute per square foot is realistic.

There is no reason for supplying large amounts of air to a large area when never more than a few people occupy that space. Outdoor air supplied should be reduced to minimum local requirements and the exhaust requirements should be balanced to maintain a slight positive pressure, retarding air infiltration and thereby reducing heat losses or heat gains. Dampers should be as airtight as possible in the closed position. The reduction of exhaust air balanced proportionally to inside requirements will reduce the load on the air-conditioning and/or heating system. To maximize efficiencies, ventilation should be scheduled so that the exhaust system operates only when it is required.

The installation of economizer enthalpy controls should be explored. These controls minimize cooling energy required by utilizing the proper amounts of outside air. (See Figures 3-6 through 3-12.) If capabilities do not exist within the plant, seek the advice of a competent consultant.

Another consideration is the installation of dry indirect evaporative cooling. This system employs an indirect evaporative cooler, with air passing through a series of tubes, cooled by air and water on the outside. Cooling can reduce energy consumption effectively within a given air-conditioning system.

The following charts and table illustrate a true economizer cycle. Inherent in this cycle are supply air reset, optimal start time and optimal chiller operation.

Figure 3-6 Economizer Cycle

- True Economizer Cycle

- Supply Air Reset

- Optimal Start Time

- Optimal Chiller Operation

Figure 3-7 True Economizer Cycle (Hardware)

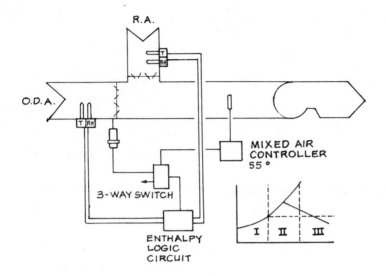

Figure 3-8 True Economizer Cycle (Computer)

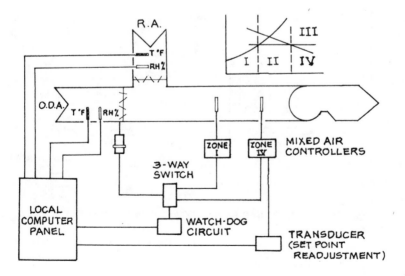

Figure 3-9 D/P-D/B Control of S/A & Supply Air Reset

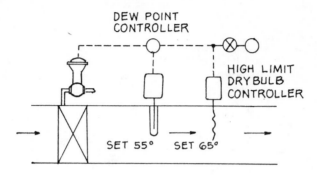

DEW POINT-DRY BULB CONTROL OF SUPPLY AIR

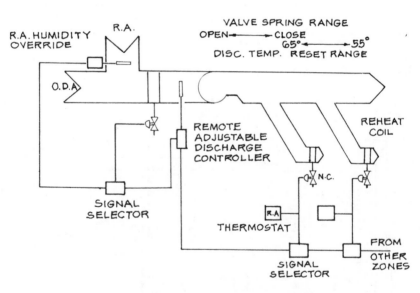

SUPPLY AIR RESET   (HARDWARE)

Table 3-1 Conventional Economizor Cycle

City	True Economizer Cycle Energy	Switchover * Temperature		
		55F Dry Bulb	60F Dry Bulb	65F Dry Bulb
Dallas	64.45	73.87	69.08	66.17
Houston	82.71	92.06	87.12	84.26
Miami	112.96	118.44	116.41	114.70
Milwaukee	29.96	38.52	34.18	31.29
New York	36.86	46.86	41.57	38.55
Phoenix	72.99	84.67	79.74	76.08
San Francisco	26.01	52.76	36.74	29.03

*O.d.A. temperature above which outdoor air damper is maintained at its minimum position.

Energy Analysis of Economizer Cycle

Average year based on 10 years of U.S. Weather Bureau data.
Figures indicate one year's energy use in millions of BTUs per 1,000 cfm, 20 percent minimum O.d.A.-return air 75° F & 50 percent R.H., 12 hr/day, 7 day/week operation (6 a.m. to 6 p.m.).

Figure 3-10 Supply Air Reset (Computer)

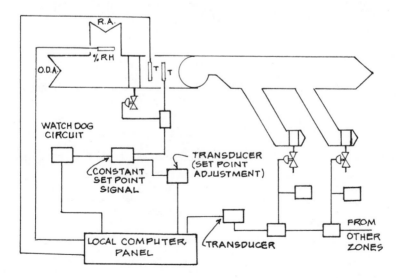

Summary of Savings from Supply Air Reset Program

(Per 1 Year, 100,000 cfm, 5-Day-Per-Week Operation, ° F Reset)

Summer	**Milwaukee**	**Denver**
Cooling savings $359.40		$169.31
Reheat savings 343.14		161.65

Winter		
Reheat savings 129.30		138.78
Humidification savings 84.81		117.29
Combined savings $916.65		$587.03

NOTE: The following formulae were used in calculating the above figures:

Cooling Savings

$$= \Delta H \times \text{lbs/hr flow} \times \text{hrs of occurrence}$$
$$\div \frac{BTU}{\text{ton-hr}} \times \text{\$ per ton-hr (1)}$$

Reheat Savings

$$= \Delta H \times \text{lbs/hr flow} \times \text{hrs of occurrence}$$
$$\times \text{\$ per lbs. of steam (2)}$$

Humidification Savings

$$= \Delta \text{ sp. humidity} \times \text{lbs/hr flow} \times \text{hrs of occurrence} \times \text{\$ per lbs of steam (2)}$$

(1) $0.0298/ton-hr.
(2) $2.5 per 1,000 lbs of steam

Figure 3-11 Indirect Evaporative Cooling Performance Chart

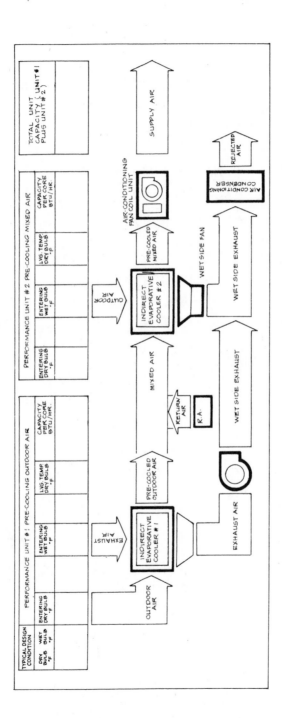

Figure 3-12 Abridgement of Psychrometric Chart

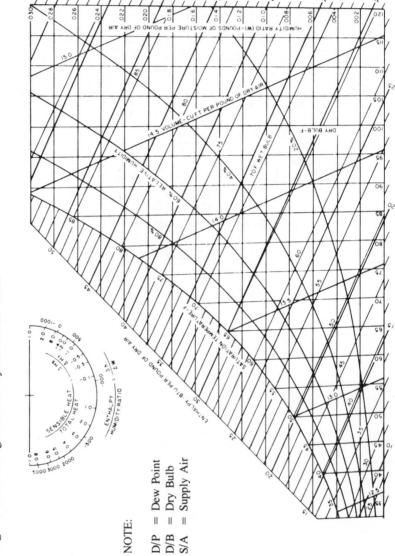

NOTE:

D/P = Dew Point
D/B = Dry Bulb
S/A = Supply Air

Chapter 4

Maintenance Procedures and Modifications

GUIDELINES FOR MAINTENANCE

The manager of an organization can utilize a number of practical tests to determine the type of improvements needed for energy conservation throughout the plant. The importance of maintenance to an energy management program cannot be over-emphasized. Effective maintenance not only will help ensure efficient operation of equipment and systems, it also will help prolong the usable life of equipment.

The maintenance guidelines presented here should be performed to bring systems at least up to efficiency. They also should be continued on a regularly scheduled basis depending on the nature of the system, frequency of operation, etc. Inspection of many of the items mentioned also will provide some idea of the effectiveness of the current maintenance program and the condition of the equipment, some of which may need adjustment, repair, or replacement.

These guidelines are general only. Wherever possible, the manufacturer of the equipment involved should be contacted to obtain literature describing the maintenance procedures suggested. Otherwise, those who regularly install such equipment or who design heating and cooling systems should be asked to prepare manuals or guidelines for their operation.

FANS, PUMPS, AND MOTORS

Proper maintenance of fans, pumps and motors can improve operating efficiency greatly and so eliminate unnecessary energy consumption. The following maintenance guidelines are suggested:

Fans

- Check for excessive noise and vibration; determine cause and correct as necessary.

- Keep fan blades clean.
- Inspect and lubricate bearings regularly.
- Inspect drive belts; adjust or replace as necessary to ensure proper operation — proper tensioning of belts is critical.
- Inspect inlet and discharge screens on fans; keep them free of dirt and debris at all times.
- Inspect fans for normal operation.

Pumps

- Check for packing wear that can cause excessive leakage; repack to avoid excessive water wastage and shaft erosion.
- Inspect bearings and drive belts for wear and binding; adjust, repair, or replace as necessary.

Motors

- Check alignment of motor to equipment driven; align and tighten as necessary.
- Check for loose connections and bad contacts on a regular basis; correct as necessary.
- Keep motors clean.
- Eliminate excessive vibrations.
- Lubricate motor and drive bearings on a regular basis to help reduce friction and excessive torque that can result in overheating and power losses.
- Replace worn bearings.
- Tighten belts and pulleys to eliminate excessive losses.
- Check for overheating; it could be an indication of a functional problem or lack of adequate ventilation.
- Balance three-phase power sources to motors; an imbalance can create inefficient motor operation and use of more energy.
- Check for overvoltage or low voltage condition on motors; correct as necessary.

AIR-HANDLING EQUIPMENT

Proper maintenance of air-handling equipment can improve its efficiency significantly. Proper maintenance guidelines include:

- Inspect ductwork for air leakage; seal all leaks by taping or caulking.
- Inspect ductwork insulation; repair or replace as necessary.

- Utilize ductwork access openings to check for any obstructions such as loose hanging insulation (in lined ducts), loose turning vanes and accessories, and closed fire dampers; adjust, repair, or replace as necessary.
- Inspect damper blades and linkages; clean, oil, and adjust them on a regular basis.
- Inspect air valves in dual duct mixing boxes to ensure full seating and minimum air leakage.
- Inspect mixing dampers for proper operation; adjust as necessary.
- Clean or replace air filters on a regular basis.
- Inspect air-heating, cooling, and dehumidification coils for cleanliness; coils can be kept clean by using a mixture of detergent and water in a high pressure (500 psig) portable cleaning unit.
- Inspect for leakage around coils or out of the casing; seal all leaks.
- Inspect all room air outlets and inlets (diffusers, registers; and grilles); keep them clean and free of all dirt and obstructions.
- Inspect air washers and evaporative air-cooling equipment for proper operation; clean damper blades and linkages if so equipped; inspect nozzles and clean as necessary.
- If electronic air cleaners are installed, check them regularly for excessive accumulations on the ionizing and grounding plate section; replace filter media if necessary; follow manufacturer's written instructions whenever adjustment or maintenance is required.
- Inspect humidifier/dehumidifier air dampers, fan parts, spray chamber, diffuser, controls, strainer, and eliminator; keep them all free of dirt, lint, and other foreign particles; clean eliminator wheel by directing a high pressure stream of water between blades.
- Adjust all VAV (variable air volume) boxes so they operate precisely to prevent overheating or overcooling, which wastes energy.
- Follow guidelines suggested for fan maintenance.

REFRIGERATION EQUIPMENT

Efficiency of refrigeration equipment can be improved considerably through following maintenance procedures. The following guidelines are suggested:

Circuit and Controls

- Inspect moisture-liquid indicator on a regular basis. If the color of the refrigerant indicates "wet," it means there is moisture in the system. This is a particularly critical problem because it can cause improper operation or costly damage. A competent mechanic should be called in to perform necessary adjustments and repairs immediately. If there are bubbles in the refrigerant flow as seen through the moisture-liquid indicator, it may indicate the system is low on refrigerant. Call in a mechanic to add refrigerant if necessary and to inspect for possible refrigerant leakage.

- Use a leak detector to check for refrigerant and oil leaks around shaft seal, sight glasses, valve bonnets, flanges, flare connections, relief valve on the condenser assembly, and at pipe joints to equipment, valves, and instrumentation.
- Inspect equipment for any visual changes such as oil spots on connections or on the floor under equipment.
- Inspect liquid line leaving the strainer. If it feels cooler than liquid line entering the strainer, it is clogged. If it is badly clogged, sweat or frost may be visible at the strainer outlet. Clean as required.
- Observe noise made by the system. Any unusual sounds could indicate a problem. Determine cause and correct.
- Establish normal operating pressures and temperatures for the system. Check all gauges frequently to ensure that design conditions are being met. Increased system pressure may be due to dirty condensers, which will decrease system efficiency. High discharge temperatures often are caused by defective or broken compressor valves.
- Inspect tension and alignment of all belts and adjust as necessary.
- Lubricate motor bearings and all moving parts, where applicable, according to manufacturer's recommendations.
- Inspect insulation on suction and liquid lines; repair as necessary.

Compressor

- Look for unusual compressor operation such as continuous running or frequent stopping and starting, either of which may indicate inefficient operation. Determine cause, and correct if necessary.
- Observe noise made by the compressor. If it seems excessively noisy, it may be a sign of a loose drive coupling or excessive vibration. Tighten compressor and motor on the base. If noise persists, call a competent mechanic.
- Check all compressor joints for leakage; seal as necessary.
- Inspect the purge for air and water leaks; seal as necessary.
- Inspect instrumentation frequently to ensure that operating oil pressure and temperature agree with manufacturer's specifications.

Air-Cooled Condenser

- Keep fan belt drive and motor properly aligned and lubricated.
- Inspect refrigeration piping connections to condenser coil for tightness; repair all leaks.

- Keep condenser coil face clean to permit proper air flow.
- Determine if hot air is being bypassed from the fan outlet to the coil inlet; if so, correct the problem.

Evaporative Condenser

- Inspect piping joints and seal all leaks.
- Remove all dirt from coil surface by washing it down with high velocity water jets or nylon brush.
- Inspect air inlet screen, spray nozzle or water distribution holes, and pump screen; clean as necessary.
- Use water treatment techniques as local water supply leaves surface deposits on the coil.
- Follow guidelines for fan and pump maintenance.

Water-Cooled Condenser

- Clean condenser shell and tubes by swabbing with a suitable brush and flushing out with clean water. Chemical cleaning also possible, although it is suggested that a water teatment company be consulted first.

Cooling Towers

- Perform chemical analysis to determine if solid concentrations are being maintained at an acceptable level.
- Check overflow pipe clearance for proper operating water level.
- Check fan by listening for any unusual noise or vibration; inspect condition of V-belt; align fan and motor as necessary.
- Follow guidelines for fan maintenance.
- Keep tower clean to minimize both air and water pressure drop.
- Clean intake strainer.
- Determine if there is air bypass from tower outlet back to inlet; if so, bypass may be reduced through addition of baffles or higher discharge stacks.
- Inspect spray filled or distributed towers for proper nozzle performance; clean nozzles as necessary.
- Inspect gravity-distributed tower for even water depth in distribution basins.
- Monitor effectiveness of any water treatment program that may be under way.

Chillers

- Keep chillers clean; inspect on a regular basis.
- Inspect for evidence of clogging; call in qualified mechanic to service equipment in accordance with manufacturer's specifications.

Absorption Equipment

- Clean strainer and seal tank on a regular basis.
- Lubricate flow valves on a regular basis.
- Follow manufacturer's instructions for proper maintenance.

Self-Contained Units

(Includes windows and through-the-wall units, heat pump, etc.)

- Clean evaporator and condenser coils.
- Keep air intake louvers, filters, and controls clean.
- Keep air flow from units unrestricted.
- Caulk openings between unit and windows or wall frames.
- Check voltage, as full power voltage is essential for proper operation.
- Follow guidelines suggested for compressor, air-cooled condenser, and fans.

HEATING EQUIPMENT

There are numerous kinds of heating systems installed in office buildings and retail stores. Certain common maintenance guidelines to improve efficiency of operation include the following:

Boilers (General)

- Inspect boilers for scale deposits, accumulation of sediment, or boiler compounds on waterside surfaces. Rear portion of boiler must be checked because it is area most susceptible to formation of scale. (Scale reduces the efficiency of the boiler and can lead to overheating of furnace, cracking of tube ends, and other problems.)
- Inspect fireside of the furnace and tubes for deposits of soot, flyash, and slag, as well as fireside refractory surface. Soot on tubes decreases heat transfer and lowers efficiency. (If boiler does not have one now, consider installation of a thermometer in vent outlet. It can save inspection time and often is more accurate than visual inspection alone.) If gas outlet temperature rises above normal, it can mean tubes need cleaning. Evidence of heavy sooting in short periods could be a signal of too much fuel and not enough air. Adjustment of air/fuel ratio is required to obtain clean burning fire.
- Inspect door gaskets; replace them if they do not provide a tight seal.
- Keep daily log of pressure, temperature, and other data obtained from instrumentation. This is the best method available to determine the need for

tube and nozzle cleaning, pressure or linkage adjustments, and related measures. Variations from normal can be spotted quickly, making possible immediate action to avoid serious trouble. Certain indicators can alert staff to problems that occur in oil-fired units. For example, an oil pressure drop may indicate a plugged strainer, faulty regulating valve, or an air leak in the suction line. An oil temperature drop can indicate temperature control malfunction or a fouled heating element. On a gas-fired unit, a drop in gas pressure can indicate a drop in the gas supply pressure or malfunctioning regulator.

- Note firing rate when log entries are made. Even a sharp rise in stack temperature does not necessarily mean poor combustion or waterside or fireside fouling. During load change, stack temperatures can vary as much as 100° F in five minutes.
- Inspect stacks; they should be free of haze; if not, a burner adjustment probably is necessary.
- Inspect linkages periodically for tightness; adjust when slippage or jerky movements are observed.
- Observe the fire when the unit shuts down; if fire does not cut off immediately, it could indicate a faulty solenoid valve; repair or replace as necessary.
- Inspect nozzles or cup of oil-fired units on a regular basis; clean as necessary.
- Check burner firing period; if it is improper, it could be a sign of faulty controls.
- Check boiler stack temperature; if it is too high (more than 150° F above steam or water temperature) clean tubes and adjust fuel burner.
- Check analysis of flue gas on a periodic basis. Check oxygen (O_2), carbon monoxide (CO), and carbon dioxide (CO_2). Oxygen should be present to not more than 1 or 2 percent. There should be no carbon monoxide.
- For a gas-fired unit, CO_2 should be present at 9 or 10 percent. For #2 oil, 11.5 to 12.8 percent; for #6 oil, 13 to 13.8 percent.
- Maintain air-to-fuel ratio properly. If there is insufficient air, fire will smoke, cause tubes to become covered with soot and carbon, and thus lower heat transfer efficiency. If too much air is used, unused air is heated by combustion and exhausted up the stack, wasting heat energy. Most fuel service companies will test your unit free of charge or for a token fee.
- Inspect all boiler insulation, refractory, brickwork, and boiler casing for hot spots and air leaks; repair and seal as necessary.
- Replace all obsolete or little-used pressure vessels.
- Examine operating procedures when more than one boiler is involved. It is far better to operate one boiler at 90 percent capacity than two at 45 percent capacity each. The more boilers used, the greater the heat loss.
- Clean mineral or corrosion buildup on gas burners.

Boilers (Fuel Oil)

- Check and repair oil leaks at pump glands, valves, or relief valves.
- Inspect oil line strainers; replace if dirty.
- Inspect oil heaters to ensure that oil temperatures are being maintained according to manufacturer's or oil supplier's recommendations.

Boilers (Coal Fired)

- Inspect coal-fired stokers, grates, and controls for efficient operation; if ashes contain an excessive amount of unburned coal, it probably is a sign of inefficient operation.

Boilers (Electric)

- Inspect electrical contacts and working parts of relays and maintain in good working order.
- Check heater elements for cleanliness; replace as necessary.
- Check controls for proper operation; adjust as necessary.

Central Furnaces, Makeup Air Heaters, and Unit Heaters

- Keep all heat exchanger surfaces clean; check air-to-fuel ratio and adjust as necessary.
- Inspect burner couplings and linkages; tighten and adjust as necessary.
- Inspect insulation and repair or replace as necessary.
- Follow guidelines suggested for fan and motor maintenance.

Radiators, Convectors, Baseboard, and Finned-Tube Units

- Inspect for obstructions in front of the unit and remove them whenever possible. Air movement in and out of convector unit must be unrestricted.
- Vent air that sometimes collects in the high points of hydronic units to enable hot water to circulate freely throughout the system; otherwise, the units will short cycle (go on and off quickly), wasting fuel.
- Keep heat transfer surfaces of radiators, convectors, baseboard, and finned-tube units clean for efficient operation.

Electric Heating

- Keep heat transfer surfaces of all electric heating units clean and unobstructed.
- Keep air movement in and out of units unobstructed.

- Inspect heating elements, controls, and, as applicable, fans on a periodic basis to ensure proper functioning
- Check reflectors on infrared heaters for proper beam direction and cleanliness.
- Determine if electric heating equipment is operating at rated voltage.
- Check controls for proper operation.

HOT AND CHILLED WATER PIPING

Proper maintenance of hot and chilled water piping will improve the efficiency of the piping system. Guidelines for effective maintenance of piping systems include:

- Inspect all controls; test for proper operation; adjust, repair, or replace as necessary; check for leakage at joints.
- Check flow measurement instrumentation for accuracy; adjust, repair, or replace as necessary.
- Inspect insulation of hot and chilled water pipes; repair or replace as necessary. Be certain to replace any insulation damaged by water. Determine source of water leakage and correct.
- Inspect strainers; clean regularly.
- Inspect heating and cooling heat exchangers. Large temperature differences may indicate air binding, clogged strainers, or excessive amounts of scale. Determine cause of condition and correct.
- Inspect vents and remove all clogs. Clogged vents retard efficient air elimination and reduce efficiency of the system.

STEAM PIPING

Proper maintenance of steam piping will prevent unnecessary wastage of steam, among other things. Effective maintenance procedures include:

- Inspect insulation of all mains, risers, branches, economizers, and condensate receiver tanks; repair or replace as necessary.
- Check automatic temperature control system and related control valves and accessory equipment to ensure they are regulating system properly in the various zones in terms of building heating needs, not system capacity.
- Inspect zone shutoff valves; all should be operable so steam going into unoccupied spaces can be shut off.
- Inspect steam traps. Their failure to operate correctly can have significant impact on overall efficiency and energy consumption of the system. Several different tests can be used to determine operations.

- Listen to the trap to determine if it is opening and closing when it should be.
- Feel pipe on the downstream side of trap. If it is excessively hot, trap probably is passing steam. This can be caused by dirt in trap, valve-off stem, excessive steam pressure, or worn trap parts (especially valve and seats). If it is moderately hot — as hot as a hot water pipe, for example — it probably is passing condensate, which it should do. If it is cold, trap is not working at all.
- Check back pressure on downstream side.
- Measure temperature of return lines with a surface pyrometer. Measure temperature drop across trap. Lack of drop indicates steam blow-through. Excessive drop indicates trap is not passing condensate. Adjust, repair, or replace all faulty traps.
- Inspect all pressure-reducing and regulating valves and related equipment; adjust, repair, or replace as necessary.
- Inspect condensate tank vents. Plumes of steam are an indication of one or more defective traps. Determine which traps are defective and adjust, repair, or replace as necessary.
- Check accuracy of recording pressure gauges and thermometers.
- Inspect pump for satisfactory operation, looking particularly for leakage at the packing glands.
- Correct accuracy of recording pressure gauges and thermometers.
- Correct sluggish or uneven circulation of steam. It usually is caused by inadequate drainage, improper venting, inadequate piping, or faulty traps and other accessory equipment.
- Correct any excessive noise that may occur in the system to provide more efficient heating and to prevent fittings from being ruptured by water hammer.
- Check vacuum return for leaks. Air drawn into the system causes unnecessary pump operation, induces corrosion, and causes the entire system to be less efficient.

GENERAL SYSTEM MODIFICATIONS

The only way to determine the efficiency of a heating/cooling system is to have the entire system tested and balanced in accordance with procedures outlined by the American Society of Heating, Refrigerating, and Air-Conditioning Engineers in its 1976 ASHRAE Handbook and Product Directory (Available from ASHRAE, 345 East 47th Street, New York, N.Y. 10017). The extent of testing required and its cost will depend primarily on the type of system involved (and therefore no cost categorization has been included here). In all cases, the survey itself should be conducted by a qualified consultant who has a thorough knowledge of heating and cooling systems, experience in testing and balancing, and the instrumentation and manpower required to collect and analyze the necessary data.

Chapter 5

Industrial Water and Potable Hot Water

INDUSTRIAL WATER

Probably no single substance in general use is more misunderstood than water. This misunderstanding is widespread among top executives of establishments using water in various industrial applications. Companies using industrial water are concerned primarily with the properties that most directly affect the processes at their plants. The research chemist may be looking for some rare phenomena such as polywater (polymerized water) that has no basic relationship with water in its familiar form; the ecologist is dedicated to the elimination of pollution, sometimes with little regard for resultant side effects; the plant operator is concerned with meeting cooling and heating loads and keeping his equipment functioning efficiently; the top executive often is concerned more with dollars in the ''bottom line'' sense than with how effectively money is being spent; and the water-treating engineer endeavors to prevent corrosion, scale, and organic growths while still satisfying the economic, environmental, and operating requirements.

With this broad range of interests and requirements in water quality it is understandable that each one represents a different viewpoint and approach to the basic problem of providing a heat exchange medium that will do a satisfactory job. It is difficult to avoid conflicts among all of the interests of economy, ecology; and operating efficiencies and still provide a single simple treating procedure. Generalizations are dangerous yet difficult to avoid when discussing such a complex topic as water quality. One valid generalization is that the potential for problems in the use of water never should be ignored and every effort should be made to avoid those that could be prevented or minimized by proper original system design and investigation of supply water quality, or improved by adequate and consistent water treatment.

Water is used for many functions in the overall chilled water system and varying qualities will be required for each one. The specific use discussed here is for central plant chilled water systems. Water functions as the heat transfer medium for transporting a cooling capability or as a heating medium. It is used for condensing refrigerants and thus removing the heat picked up by the chilled water throughout its distribution system. This heat finally is dissipated through a cooling tower, where water is used to reduce the temperature. (In certain instances the temperature of the water is reduced by air cooling.) The tower water is cooled by evaporation, causing a reconcentration of the salts or dissolved solids present in the raw makeup water. To hold these salts to a tolerable level, it is necessary to conduct a bleed or blowdown of the concentrated water, which is a waste water.

The Environmental Protection Agency (EPA) has been working to establish standards for waste water quality. The timetable for complying with the new standards was expected to strain the ingenuity, budgets, and technical capabilities of everyone involved with industrial wastes. It is not too soon to recognize the problems these criteria and timetables will impose. Too many in top management lack the technical training and understanding to realize they have potential water quality problems until there is a serious breakdown or a government agency issues a cease and desist order.

Until recently the restraints on industry for quality of industrial wastes have been mild, particularly for those operating in areas where there are ocean outfalls for sewage disposal. Some industries and large water users have been located on rivers where sufficient dilution of wastes has allowed them to dispose of matter that otherwise might have been prohibited and that most assuredly will be prohibited in the future. In addition to chemical pollution, thermal pollution of rivers has become a serious concern. Where large river flows were available, once-through cooling was common practice; however, with the influx of new users, each increment of temperature rise is creating problems downstream.

With these general problems established, it is imperative that variables in water quality between different locations be recognized. Any attempt to transpose a specific treatment and procedure from one location to another may lead to serious problems unless all of the variables are accounted for carefully. Some 12 to 15 major constituents generally are included in a fairly complete water analysis. Each of these constituents may be present in a wide range of concentrations, thus leading to an almost infinite variability among different water sources. It is not difficult to see how these variables can affect the selection of treating chemicals or procedures. For example, many waters are known for their high susceptibility to scaling as the result of large concentrations of water hardeners such as calcium carbonate. Other waters are more prone to cause corrosion. Still others present problems from organic material and growths of various kinds.

In all cooling tower operations, a controlled bleed or blowdown must be maintained to hold reconcentration levels to tolerable limits. The use of acid coupled with corrosion inhibitors has been standard practice for maintaining large cooling towers. The acid generally is sulfuric, which adds sulfates to the water in the tower. A large percentage of the inhibitors used in combination with the acid has involved chromate compounds. The EPA has been establishing levels for chromate disposal that are expected to eliminate this material as a corrosion inhibitor. The EPA regulations were to be enforced by agencies such as local or state water pollution boards or health departments. A grace period was to be provided for those located within the areas of ocean outfall disposal, while inland disposal quality control was expected to be tightened quickly.

Numerous unique problems may arise from the characteristics of different types of supply water. For example, when water high in carbon dioxide (usually from a well) is exposed to air, it becomes highly corrosive to copper. This is one of the many characteristics that should be determined before water is put to industrial use. A problem such as this can be corrected by pretreatment.

Water-treating chemical companies are researching these problems constantly to provide materials that will minimize the problems. Many industries with extensive water uses and problems also are doing independent research to develop treating procedures that will not pollute and will be more effective in minimizing corrosion and scale. New materials such as the polyphosphonates and aminomethylphosphonates, polyalcrylates, and other polymers are showing merit but need more field testing to be fully proved as all-around solutions. New products often are promoted too vigorously on the basis of a few unique tests. For example, it is not known widely that zinc added to many of these materials or to chromate provides a synergistic effect that often greatly enhances the corrosion inhibition of these compounds.

Generally speaking, whether a refrigeration system for air conditioning or industrial process cooling is involved, the problems of water quality for the heat exchange system will be substantially the same. The advantage of central plants is that they are close-coupled except for the distribution of the chilled or heated water. This is carried in closed systems where it has no chance of being contaminated by air or other external substances. Separate treating procedures are required for the chill water, cooling tower water, boilers, and condensate. For each separate system, a separate central treating location can be used. To assure proper treatment when systems are scattered over wide areas, each would require the same treating equipment as is needed for the central system and each would have to have its own analysis.

Most plants use internal chemical treatments and inject them into the system. An exception to this is plants using zeolite softening equipment for external pretreamment of their supply water. Other external processes such as lime and lime-soda have been limited primarily to large industrial or municipal water treating requirements.

As an offshoot of the research to desalinize sea water, two processes have been developed that hold promise for pretreatment in the systems under discussion. Reverse osmosis and electrodialysis are capable of doing an effective job but at a price that is not yet justified fully over internal treatment. In special cases they may well be the only acceptable solutions. These are membrane processes. The reverse osmosis process in essence forces pure water through a membrane under significantly high pressure, about 400 psi. Electrodialysis, using electrically charged and polarized plates and membranes, attracts the ions to be rejected through the membranes and leaves water reduced materially in ionic contaminants. The waste stream of the reverse osmosis is upstream of the membranes, which act as a filter, whereas the electrodyalysis waste stream is downstream of the membranes. Initial costs for use with normal municipal supplies are substantial, especially in light of the fact that these systems neither add to nor subtract from the final combined waste flows.

One characteristic universal to all waters is that dissolved oxygen is the major culprit causing corrosion. Every effort should be made to keep closed recirculating systems tight and avoid unnecessary makeup water laden with oxygen. Closed systems generally use oxygen scavengers because, once the oxygen is eliminated, the chance for corrosion is nil. Cooling towers, where the water is being aerated continually, present another problem in that there is no practical way of removing oxygen. So far, chemical treatment is the only practical means of controlling scale, corrosion, algae, and other organic growths. Experience dictactes that if sunlight can be eliminated from direct contact with the tower water, algae will be minimized. This can be accomplished by covering the distribution basins at the top of the tower.

Beyond all after-the-fact solutions there is one practical answer to the corrosion problem in a cooling tower system: Build the tower and all of the components of the piping and control system with corrosion-resistant materials. Components of a corrosion-resistant system include fabrication with cement-asbestos or ceramics, tower basins of cement, piping of cement-asbestos or plastic, and fittings of nonferrous alloys. Where straight ferrous metals are used, they should be coated before assembly with high-solids epoxy. This type of construction is more expensive initially but over the long-range projected life of the equipment might prove to be the most economical way to prevent corrosion-related problems.

The balancing of water systems is imperative for economic operation. Water, corrosive in nature, will deteriorate systems if not controlled. Alkaline water with high solids can impact calcium carbonate on piping and tubes used specifically for heat transfer. Scale within a cooling system and condenser section of 1/32'' can increase energy consumption by as much as 10 percent. The heavier the scale, the more energy consumed.

To get a completely objective analysis and recommendation of plant/water problems, it probably will be necessary to hire a consulting engineer. He will provide information and insights that will help in the selection of the products best suited for the problems at hand, including their costs.

POTABLE HOT WATER

Heat losses from uninsulated hot water system distribution piping can be substantial. The degree of heat loss depends on the temperature differential between pipe and ambient air, pipe size, and length of system piping.

Exposed piping in basements and equipment rooms is relatively simple to insulate. Piping in ceiling spaces may be readily accessible by removing ceiling panels. Ideally, the entire piping system should be insulated, but inaccessible portions may be left bare providing they are a small percentage of the total, as this will have little effect on the total possible savings.

Installation costs vary in each building and should be estimated by an insulation contractor. For order of magnitude, use the installed costs for sectional insulation with jacket listed in Table 5-1.

Table 5-1 Sectional Insulation

Pipe Size	Price Per Lineal Foot Installed	
	Insulation Thickness	
	1''	1½''
	$	$
½ ''	1.55	2.25
¾ ''	1.60	2.35
1 ''	1.75	2.40
1¼''	1.90	2.60
1½''	2.05	2.75
2 ''	2.20	2.85
2½''	2.35	2.95
3 ''	2.50	3.05
4 ''	2.65	3.25
5 ''	2.80	3.50
6 ''	2.95	3.75
8 ''	3.60	4.50
10 ''	4.30	5.50
12 ''	4.80	6.40
14 ''	5.50	7.50
16 ''	6.50	8.20
18 ''	7.20	8.80
20 ''	8.75	9.50
24 ''	9.75	9.80

For costing purposes, add to total lineal feet of piping three lineal feet for each fitting or pair of flanges to be insulated. As an example of how to estimate savings a separate oil fired water heater with an efficiency of 70 percent provides the domestic hot water. Oil saved totals 5,485 gallons; at $.47 per gallon, this equals $2,578 per year.

Order of magnitude costs are:

$$
\begin{aligned}
125 \times \$2.35 &= \$293.75 \\
30 \times \$2.05 &= \$\ 61.50 \\
350 \times \$1.90 &= \$665.00 \\
150 \times \$1.75 &= \$262.50 \\
\hline
&\ \ \$1,282.75
\end{aligned}
$$

An investment of approximately $1,300 will save $2,600 a year. No further economic analysis is necessary.

Insulate Storage Tanks

The loss of heat from the domestic storage tank must be offset continuously by the addition of heat to maintain a ready supply of hot water. This heat loss occurs 24 hours a day whether or not the building is occupied.

Cover storage tanks with a minimum of three inches of insulation. Insulate bare tanks and apply additional insulation to tanks having less than three inches. Replace or repair all missing or torn insulation as required.

Check applicable codes to determine acceptability of various insulation materials.

Obtain quotations as a basis for accurate cost assessments, such as a per-square-foot estimate for surface area of tanks.

Calculate energy savings by determining the heat lost from the tank before and after insulation. Assume water temperature and ambient air temperature are constant. Multiply the savings in BTU/hr. $\times$ the constant 8,760 to obtain the total yearly energy savings due to insulation.

Replace Central System by Local Heating Units

Commercial hot water systems frequently require hot water for short periods of heavy use at various locations within the building. It often is more efficient to provide water heaters close to the usage points rather than use central generation and long runs of hot water piping.

Analyze hot water use within the building to determine the patterns of demand and thereby decide whether installation of local units is advantageous. Estimate the energy losses of the existing system and calculate the savings to be derived by installing local units. Compare the energy saved in the sum of reduced distribution losses, and the increase in the average generation efficiency of local units, to a central system.

Before installing local hot water generation units, review codes to see if any restrictions are required on locations or modifications to the building such as fire walls around fuel burning equipment.

Install Temperature Boosters

When multiple temperature requirements are met by a central domestic hot water system, the minimum generation temperature is determined by the maximum usage temperature. Lower temperatures are attained by mixing with cold water at the tap. Where the majority of hot water usage is at the lower temperatures and higher temperatures are required at only a few specific locations, install booster heaters or separate heaters for high temperatures.

Install Separate Boiler for Summer Hot Water

In many buildings, the heating system boilers provide primary heat for the domestic hot water system. While this is satisfactory during the heating season when the boiler is firing at high efficiency, demand for boiler heat in summer probably will be limited to hot water generation only. Operating large heating boilers at light loads to provide domestic hot water results in low boiler efficiency.

To reduce energy losses due to low boiler efficiency in summer, install a separate water heater or boiler, sized for the hot water demand. Shut down the heating boiler in the summer and generate hot water at improved efficiency.

Obtain quotations from a local contractor to determine the cost of this modification.

To determine the energy savings, analyze the system operating characteristics. Items investigated should include operating efficiencies of existing equipment during warm and cold months and hot water demand relationship to boiler capacity. Calculate savings by comparing the operating costs of both arrangements.

Hot Gas Heat Exchanger

A typical refrigeration machine with a water-cooled condenser rejects approximately 15,000 BTU/hr. for each 12,000 BTU/hr. of refrigeration. An air-cooled condenser rejects up to 17,000 BTU/hr. Up to 5,000 BTU/hr. of the heat rejected from either system can be recaptured. To recover heat of compression, install a heat exchanger in the hot gas line between the compressor and condenser of the chiller. A typical arrangement in conjunction with a domestic hot water system is shown in Figure 5-1. Hot gas temperature depends on head pressure but is usually in the order of 120 - 130° F.

Figure 5-1 Hot Gas Heat Exchanger

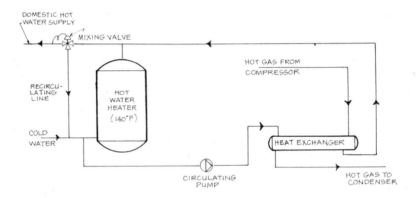

Cold water is circulated through the heat exchanger by the circulating pump. When hot water is not being used, water is pumped back through the heat exchanger through the recirculating line. When hot water is needed, it is fed from either the heat exchanger, the hot water heater, or both. A mixing valve is provided to maintain the desired temperature.

Hot Drain Heat Exchanger

In many cases, buildings with kitchens, laundries, and other service facilities that utilize large quantities of hot water, discharge hot waste water to drains. By installing a heat exchanger, heat can be recaptured and used to preheat domestic hot water.

In general, it is economical to preheat water from 50° F to 105° F without excessive equipment cost. The hot water at 105° F then can be fed into the domestic hot water tank for further heating to utilization temperature, if required. The heat reclamation system saves the heat required to raise water from 50° F to 105° F that otherwise would have been provided by other heat sources.

For example, in a laundromat the average daily wash load is 2,000 lbs. Water usage is 2.5 gallons of 170° F water and 1.0 gallon of 50° F water per lb. of wash load. The waste water discharged to drain is 3.5 gallons of 136° F water per lb. of wash load.

Assuming that the laundromat operates 365 days per year and that domestic hot water is generated by an electric hot water heater, the yearly saving is 553 × 10 BTU. At 90 percent efficiency and $0.03/kwh, this represents $5,400 cost saving a year, based on preheating the water by heat reclamation.

Figure 5-2 Schematic of Laundry and Kitchen Hot Water Heat Recovery System

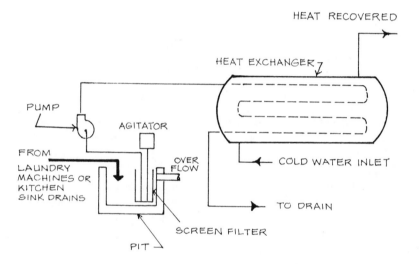

Hot Condensate Heat Exchanger

The condensate return portions of many steam systems exhaust large quantities of heat in the form of flash steam when the hot condensate is reduced to atmospheric pressure in the condensate receiver. Recover waste heat by installing a heat exchanger in the condensate return main before the receiver to reduce condensate temperature to approximately 180° F. Use the heat revovered to preheat water as required. Figure 5-3 shows the schematic installation of a hot condensate heat exchanger.

Figure 5-3 Hot Condensate Heat Exchanger

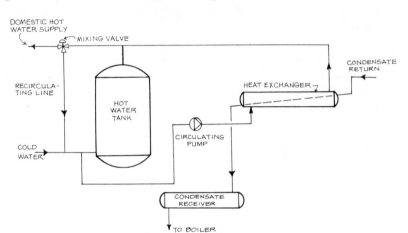

The quantity of heat recovered depends on the pressure and temperature characteristics of the boiler. For example: Steam has a condensate return volume of 6 gpm at 260° F. A heat exchanger is installed to reduce the condensate temperature from 260° F to 180° F and the quantity of heat recovered is 240 × 13^3 BTU/hr. The average heat input required for the generation of domestic hot water is 1.1×10^6 BTU/day. In this instance the entire domestic hot water load can be met the major part of the year by the heat recovered in the hot condensate heat exchanger. The hot water system represents another key area for energy reduction. By their very nature, hot water systems can be energy wasteful. The following are guidelines that can reduce energy consumption and reduce operating costs.

GUIDELINES TO REDUCE ENERGY USED FOR DOMESTIC HOT WATER SYSTEMS

Reduce Consumption of Domestic Hot Water

a. Reduce cold water pressure with an automatic control valve, but do not set lower than pressure needed to flush toilets on top floors or for pressures required for fire protection.
b. Take measures to reduce consumption when usage rates exceed 1½ gpd/person in offices and 1 gpd/person in stores or religious buildings.
c. Install flow restrictors in the supply branch to groups of taps when existing faucets have flow rates greater than 1½ gpm.

d. Install spray type faucets that permit flow of ¼ gpm instead of 2 or 3 gpm.
e. For new additions, install foot-operated peddle valves.
f. Install self-closing faucets on hot water taps.
g. Install a flow meter on the cold water line supplying the water heater when hot water consumption is more than 3,000 gallons per day.
h. Replace obsolete kitchen equipment such as dishwashers with types that have minimal water requirements.

Reduce Temperature of Domestic Hot Water

a. Lower water temperature when it exceeds 125° F.
b. Install an electric, gas, or small oil-fired boiler to boost temperature for kitchens, laundries, or processes requiring higher temperatures when the majority of domestic hot water usage is at lower temperatures. Consult code authorities for minimum allowable temperatures for commercial kitchen use.
c. Install a seven-day timer clock to deactivate water heaters to reduce temperatures further during unoccupied periods.

Reduce Thermal Loss from Hot Water Piping, Appurtenances, Storage Tanks, and Hot Water Generators

a. Insulate hot water storage tanks when insulation is less than the equivalent of 3″ fiberglass and/or when insulation is in need of repair.
b. De-energize hot water circulating pumps to reduce heat loss from piping within the building. Where extremely long runs require that the circulator be in operation during occupied periods, install a seven-day timer to shut off pump automatically during unoccupied periods.
c. Insulate exterior jacket of tankless or tank heaters that are not immersed in the hot water boiler or hot water tank.
d. Locate hot water tank as close to the load as possible when installing new storage tanks or making major modifications to the building.
e. Insulate hot water piping whenever there is less than the equivalent of 1″ fiberglass.

Increase Efficiency of Hot Water Generator and Equipment or Install More Efficient Equipment

a. Replace gas pilots with electrical ignition.
b. Install storage water heater for summer use when existing space heating boiler is used for hot water generation, when there is little or no demand for steam or hot water during the summer and when hot water accounts for 20 percent or more of the load.

   c. Replace electric water heaters with heat pumps to improve the coefficient of performance from 1 to approximately 3. Use hot drain water as a heat source for the heat pump, or use an air-to-water heat pump.

   d. Repipe hot water storage tanks if cold water makeup supply is connected to the upper half of the tank and/or if the hot water outlet from the tank is in the lower portion of the tank.

   e. Provide oil- or gas-fired water heaters or heat pumps rather than electric resistance heating when hot water demands are increased due to expansion or change in occupancy. If heating boiler has sufficient capacity and is in operation year-round for air conditioning as well, install a tank or tankless heater in place of separate hot water generator. If the additional requirements for hot water are to serve facilities remote from the boiler and usage is small, install separate heater in the cold water line directly at the fixtures rather than serving them from a central system.

   f. Install local hot water heating units when domestic hot water usage points are concentrated in areas distant from the central generation and storage point.

Use Waste Heat from Other Processes to Preheat Domestic Hot Water

   a. Install hot gas heat exchanger in air-conditioning refrigeration compressor/ systems or commercial refrigeration equipment to heat hot water.

   b. Install heat exchanger and utilize waste heat from the engine where diesel or gas engines are in use.

   c. Utilize waste heat from incinerator to generate hot water where incinerators are handling more than one ton of solid waste per day.

   d. Install heat exchanger in hot water drains from kitchens and laundries where flow exceeds 2,000 gallons per day.

   e. Use hot condenser water from refrigeration systems to preheat domestic hot water.

   f. Install heat exchanger in condensate lines from steam equipment to preheat domestic hot water.

   g. Install heat pipe or heat exchanger to extract heat from boiler breechings to preheat hot water.

Chapter 6

Industrial Pumping Equipment and Instrumentation

INDUSTRIAL PUMPING

Energy transport losses are inherent in the design of any distribution system. However, they can be minimized by reducing the flow rate through the system and eliminating obstacles causing resistance to flow.

Several options are available for reducing distribution losses. Minimizing building loads will allow flow rates to be reduced. To avoid repetitive actions, every effort should be made first to reduce building heating and cooling loads.

The energy required to distribute air and water does not help to reduce the building load. It actually is counter-productive in cooling because the heat equivalent of the power input must be removed by the refrigeration equipment.

The characteristic of any given system is predetermined by the length and size of pipes or ducts and size and shape of fittings (bends, tees). The resistance to flow is a function of velocity (or flow rate) and the system characteristic. With any change in flow rate, the system resistance will change (according to the laws of fluid flow) and the operating point will move accordingly on the system characteristic curve. (See Figure 6-1.)

As shown in Figure 6-2, modification of a piping system characteristic will change the operating point from A to B along the pump curve and result in an increased flow rate at a lower resistance. To reduce the flow rate back to its initial point, it is necessary to reduce the pump speed so the new pump curve intersects the new characteristic curve at point C, giving the desired flow rate. These principles apply equally to fan/duct systems and pump/pipe systems.

When changing flow rates and system resistance, implement the adjustments and modifications in the following logical sequence:

Figure 6-1 Heat Flow Rate

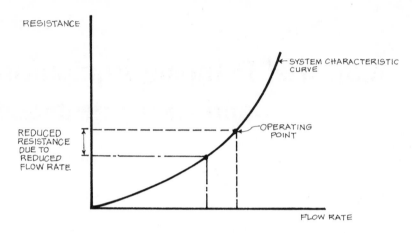

For instance, a piping system with a flow rate of 2,000 gpm at 150 ft. resistance changed to 1,800 gpm will have a new resistance of:

$$150 \times \left(\frac{1,800}{2,000}\right)^2 = 121.5 \text{ ft.}$$

Figure 6-2 Modification of Heat Flow Rate

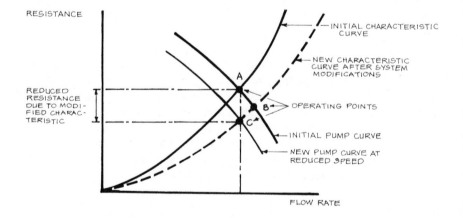

a. Measure initial flow rates and resistance and construct the system characteristic curve.
b. Reduce loads and calculate new reduced flow rate required to meet them.
c. Modify system to reduce resistance to flow, and from measurements construct the new system characteristic curve.
d. Determine from the new characteristic curve the resistance at the new reduced flow rate.
e. Reduce fan or pump speed or reduce pump impeller diameter so new fan or pump curve crosses the new system characteristic curve at the desired operating point.

Reduce Resistance to Flow in Piping Systems

The resistance to flow in a piping system is the sum of the resistance of all its individual parts in the index circuit. Some of the parts cannot be modified easily, but others are candidates for reducing resistance.

Heat exchangers have a high resistance to flow and are prone to fouling by scale deposits and dirt, which further increases flow resistance. To correct this condition, disassemble the heat exchanger, remove scale mechanically or chemically, and flush out the tubes and shell. Institute a maintenance and water treatment program for heat exchangers based on regular observations of pressure drop and temperature differentials.

Over long periods of time, formation of scale deposits that occurs throughout the system may result in radically increased resistance. Determine by inspection whether scaling exists and rectify as necessary by chemical cleaning. There are many specialized contractors who are experts in this work.

Replace high-resistance elements of filters and strainers with low-resistance elements or baskets.

Pumps must develop sufficient head to overcome the resistance to flow through the longest or index circuit, even though this head may exceed the requirements of the other subcircuits. If the index circuit resistance grossly exceeds resistances of the other circuits and cannot be reduced further, consider the installation of a small booster pump for the index circuit only and reduce the head of the main pump accordingly. (See Figure 6-3.)

Figure 6-3 Booster Pump Insertion in System

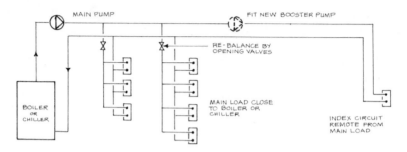

Formerly, when existing piping systems were put into operation, the installing contractor usually balanced flows by trial and error, often closing balancing valves to a greater extent than necessary and imposing extra head on the pump. Designed safety margins often result in oversized pumps. The excess head is absorbed by closing down the valve on the pump discharge. To reduce resistance to flow, rebalance the system by first opening fully the balancing valve on the index circuit and the pump(s') valve or removing any orifice plates from the pump circuit. Then adjust each circuit balancing valve to achieve proportional flow rates, starting with the next longest circuit and progressing to the shortest circuit. This process is trial and error — each valve adjustment will affect flow rates in circuits already adjusted, but two or three successive adjustments of the whole system will provide good balance. When all available options to reduce system resistance have been exercised, measure or calculate the new system characteristic and determine the new operating point, taking into account the previous steps to reduce the required flow rate.

Reduce Water Flow Rates

When the heating or cooling load actually is less than original design or is reduced further, the flow rate through the hot water or chilled water systems may be reduced proportionally, provided the supply temperature set point is retained at its initial level.

Reduce the pump speed to decrease the water flow rate through a piping system. Derive the appropriate reduction in pump speed to achieve any given reduction in flow rate from the pump curve (obtain from manufacturer). The reduction in pump speed will be approximately in proportion to the change of flow rate.

To reduce the speed of indirect drive pumps, change the size of the motor sheave. To reduce the speed of direct drive pumps, exchange the drive motor for one of lower speed. (This is possible only within the range of commercially available motor speeds and a compromise may be necessary).

If it is not possible or economically feasible to change the speed of a pump, reduce the flow rate by changing the impeller size. Substitute one of smaller diameter or skim down the existing impeller. Seek manufacturer's recommendations for each specific application.

Reducing the water flow rate through a piping system also will reduce the resistance in accordance with the laws of fluid flow. To determine the new system resistance, use the following formula:

New system resistance = Initial system resistance $\times$

$$\left(\frac{\text{New flow rate}}{\text{Initial flow rate}} \right)^2$$

To determine energy consumed by pumps, refer to Figures 6-4 and 6-5. Select the appropriate graph and enter at the initial flow rate. Following the direction of the example line, intersect with the appropriate points for initial system resistance and hours of operation.

Read out the yearly energy used in BTU/year. Repeat this procedure using the new flow rate and new system resistance. Subtract the yearly energy used at the reduced flow rate from the yearly energy used at the initial flow rate. To convert BTU/year, divide by 3,413 to obtain yearly savings.

The cost of achieving these savings will be the sum of the costs for the individual steps taken. The cost of fitting new pump sheaves will vary with each individual case but, for order of magnitude, will be approximately $100. The cost of replacing the impeller or machining it down in diameter should be obtained from the pump manufacturer.

Figure 6-4 Engineering Data: This figure is based on the Standard Pump Formula:

$$\text{Brake horsepower} = \frac{\text{lbs/min} \times \text{Feet head}}{33,000 \text{ ft/lbs} \times \text{Pump efficiency}}$$

Pump efficiency was assumed to be an average of 70 percent. Brake horsepower was converted to BTUs by the factor 2,544.43 BTU/horsepower hour.

The upper half of the graph proportions the hours of pump operation per year from 100 percent (8,760 hr.), down to 11 percent (1,000 hr.).

Figure 6-4 Centrifugal Pumps Up to 500 GPM

(Yearly Energy Consumed)

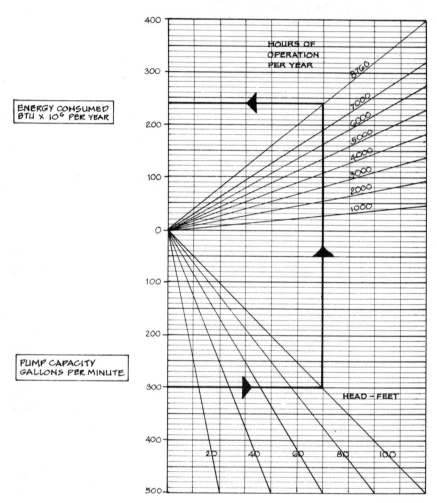

Figure 6-5 Centrifugal Pumps Up to 2,000 GPM

(Yearly Energy Consumed)

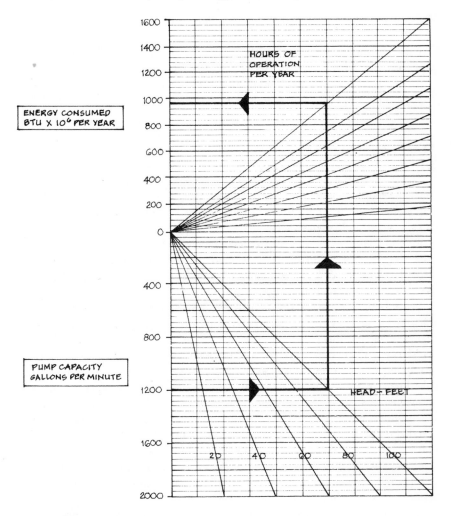

Figure 6-5 Engineering Data: This figure is based on the Standard Pump Formula:

$$\text{Brake horsepower} = \frac{\text{lbs/min} \quad \times \text{ Feet head}}{33,000 \text{ ft/lbs} \times \text{ Pump efficiency}}$$

Pump efficiency was assumed to be an average of 70 percent. Brake horsepower was converted to BTUs by the factor 3,544.43 BTU/horsepower hour.

The upper half of the graph proportions the hours of pump operation per year from 100 percent (8,760 hr.), down to 11 percent (1,000 hr.).

INSTRUMENTATION

The purpose of a central control system is to provide the physical plant chief engineer and the energy management team with a management tool to conduct constant surveillance of the building and to make the most efficient use of physical plant systems and personnel.

Central control systems are applicable only to the larger, more complex buildings or groups of buildings. They should be considered where net floor area exceeds 100,000 square feet and where energy use is high due to extended occupancy or use of special equipment.

Central control systems vary from the relatively simple type designed to perform a few functions to progressively more complex types performing more and more complicated functions.

When a central control system is selected, it should be tailored to the present requirements of the building and the systems to be controlled. If money for investment is limited, attention should be given first to those functions that will show the quickest and greatest return in energy saved. However, an overview of the central control system capacity should be made at the time of the initial installation and provisions included for expanding the system's hardware and software capacity when investment money permits. For instance, a basic system composed of a control console and central processor unit with one single printer output can be programmed to perform start/stop and simple reset functions and to report alarms. Optimization of system operation and load shedding in this case would be achieved manually at the central console based on decisions made by the operating staff and achieved by overriding the programmed start/stop times and reset points for the various systems.

Hardware options that could be added to this initial system include an extra printer to act as a backup and to handle alarm signals separately from system status reports, a cathode ray tube output to show in English the status of selected systems and current modifications made to the normal operating program, a bulk storage memory unit to increase the initial core memory included in the central processing unit so that more complex programs can be used and a data file of operating conditions built up, a high speed tape input/output device to enable new programs to be fed into the computer and to extract from the computer banks historical data of systems operation, and complex custom written software programs to optimize the operation of all systems to achieve automatic load shedding on a selected rolling priority basis.

A central control system should be selected to:

a. Monitor all systems for off-normal conditions.
b. Monitor all fire alarm and security devices.
c. Monitor operating conditions of all systems and reschedule set points to optimize energy use.
d. Monitor on a continuous basis selected portions of any system and store this information in bulk memory for later retrieval and use in the energy management program to indicate changes or modifications in approach that should be made.
e. Limit peak electrical demand values by predicting trends of loads and shedding nonessential services according to programmed priorities.
f. Optimize the operation of all systems so that the maximum effect is obtained for the minimum expenditure of energy.
g. Optimize maintenance tasks to achieve maximum equipment life for minimum labor costs.
h. Provide inventory control of spare parts, materials, and tools used for maintenance.

By judicious use of these functions, the engineering staff will be able to operate all systems from the central console and will have minute-by-minute control of the operation. Any critical alarm in the physical plant will be reported automatically at the console. The operator then will be able to scan the system in alarm, analyze the fault, and dispatch the correct maintenance person to make repairs. Maintenance alarm summaries will be available on daily printouts or on demand as needs dictate. These maintenance alarms will allow work scheduling and maximum use to be made of maintenance personnel.

System Types

Proprietary central computerized control systems are marketed by each of the major temperature control manufacturers. These systems all have common features and can accomplish a similar range of tasks. Each manufacturer, however, uses coding and computer languages that are unique to that company's system and cannot be decoded by any other system. Once a basic system has been selected and installed, all subsequent additions must be obtained from the original manufacturer. Therefore, it is important when selecting a system to investigate thoroughly its expansion potential in relation to future requirements.

Central computer control systems usually are modular in design and additional hardware and software can be added later provided expansion capability has been included in the system. (See Figure 6-6)

Figure 6-6 Typical Central Control System

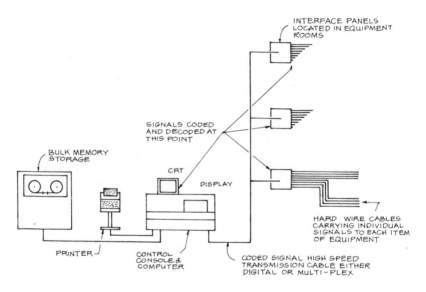

Each manufacturer's system consists of standard "off the shelf" hardware but the application always is tailored to the specific project. Basically, any system is composed of four major parts:

1. Interface panels located at strategic points throughout the building (usually equipment rooms). These panels form the focal point of all signals to and from a particular area.
2. Transmission system between the central console and all interface panels. This transmission system can be a simple single-core cable for digital transmission or multicore cable for multiplex transmission.
3. Central control console and associated hardware located in a control room (usually close to boiler room and chief engineer's office). The console, computer, and associated hardware form the point at which the operator enters all instructions and retrieves all data. In addition, the computer generates routine instructions according to the program contents.

4. Software (program) generated by the manufacturer in conjunction with the prospective user. Software input via magnetic tapes, paper tapes, or cards contains that basic operating instructions for the computer and is stored in the form of "bits" of information either in core memory or in bulk memory (cheaper than core for large quantities of data).

Selecting a Central Control System

To obtain maximum benefit from a central console system it is necessary to measure, monitor, and control many different items of equipment.

The central control system must be interfaced with the existing equipment to obtain this information in rational form and to exert its control functions.

Binary signals are used as instructions to start and stop equipment, open and close valves and dampers, etc. They are used also for retrieving data such as on/off, open/closed, etc.

Analog signals are used as instructions to raise or lower temperature set points, adjust damper positions, raise or lower pressure set points, etc. They are used also to retrieve data such as temperature, pressure, humidity, etc.

To select a central control system, it is necessary first to assemble a complete list of all desired interface points under the binary and analog categories and to arrange them in groups served by individual interface panels.

Depending on the type of existing controls, motor starters, contactors, etc., modifications and additions may be required to allow satisfactory interface. For instance, if a motor starter does not have spare auxiliary contacts, a relay must be added to the control circuit.

Signal Transmission

To make effective use of the computer's capabilities, information between the computer console and the interface panels must be transmitted at high speed in a format handled easily by the computer. The most convenient method of transmission is digital, where information is represented by pulses arranged serially and transmitted through a single core conductor. Digital cable commonly is coaxial, although some systems use a twisted pair of insulated wires. Voice intercom is carried over a separate screened cable.

Multiplex cable is multicore, signals being multiplexed and transmitted in parallel. Typically, the cable will consist of 50 to 100 separate wires and have an overall diameter of 1 inch. It is more difficult to install this cable than coaxial, particularly in existing buildings where empty conduits and throughways may not be available.

Some systems still use multiplex cable on the ground that analog signals can be transmitted in unmodified form, whereas for digital transmission they must be converted from analog to digital form, with a small loss of accuracy at the interface panel. Conversely, analog signals are sensitive to interference from ''spikes'' and other spurious signals induced by adjacent building wiring while digital signals are not subject to interference.

When selecting a central control system, the different characteristics of the transmission methods must be analyzed and a choice made based on the particular circumstances. Generaly, digital transmission systems provide more options for later additions because multiplex systems are limited in the number of points that can be connected.

Control Console and Hardware

The control console and hardware items such as printers, graphic display, cathode ray tube (CRT), and input/output keyboards are of comparable performance for any manufacturers' systems. Individual options are available on request to suit the prospective users' particular requirements. For instance, three to five different levels of access into the computer from the keyboard are available. Access would range from restriction to normal operation for the lowest level up to reprogramming for the highest level, and would be controlled by keyswitch. It is vital to protect memory from accidental erasure.

The central processor computer will be supplied with integral core memory. Capacities vary from 16,000 to 64,000 words, each word containing 16 bits of information. Core memory is expensive compared with bulk memory. If the core capacity will be exceeded, the addition of a bulk memory unit should be considered, but first make sure that the computer selected can be interfaced with external memory.

Bulk memory can be either disc or tape, both with random access. Disc memory is preferred for control applications and typically has a capacity of 192,000 words.

Software

Software or programs for common applications are available from the manufacturer. Each of the major controls manufacturers has a library of application programs that have been developed over the years. The cost of these programs is low compared to custom written software. When selecting a central control system, investigate the library of programs available as the extent and range of ability in the computer control field.

Computer programs available and their application to operations are:

Executive Program

This program is essential to the functions of the computer itself as it controls the priority of signal processing, directs information in and out of memory, controls the operation of all associated hardware, and encodes/decodes incoming and outgoing signals. An essential function is the orderly shutdown of the computer if power fails to ensure that all information is stored safely in permanent memory before all power is lost. Beware of systems that store information in volatile memory which evaporates on power failure.

Start-Stop Program

This is the simplest program and yet, with modifications, becomes the most effective energy saver, particularly when used in conjunction with other programs and routines.

In its simplest form the program is arranged to start and stop equipment according to predetermined times, allowing for weekends and holidays. The program also can open and close dampers, turn lights on and off, and reschedule temperature from occupied level to unoccupied level as all of these functions are binary in nature.

Manual override can be provided, with automatic reversion to program control at the end of an occupied or unoccupied period.

Interface with other programs allows the basic program to be overridden and equipment turned on and off to meet some other criteria.

Start sequences can be arranged in cascade to avoid simultaneous starting and the cumulative effect of inrush currents causing a high peak electrical demand.

Load Shedding Program

This is used in conjunction with the start/stop program to override normal operation and turn off equipment as a predetermined peak load value is approached. In its simplest form the program would shut down selected equipment according to the measurement of peak load. The program can be modified to include a rolling priority feature whereby equipment would be turned off on an assigned priority basis for an adjustable period of time, after which it would be returned to service and a lower priority piece of equipment turned off. The rolling priority would be activated only as a predetermined peak load value was approached, and the equipment turned off would be limited to that necessary to maintain the load below a given level.

Interface with a profile routine would allow a continuous profile of electrical demand to be made and predicted load in the immediate future to be calculated. This predicted load then would initiate the rolling priority to alter the slope of the prediction curve. The current maximum peak load experienced in the previous 11 months would be held in memory and used as the criteria for load shedding. This value would be modified constantly based on monitored information.

Optimizing Programs

These programs tend to be more sophisticated and require interface with an arithmetic routine to allow optimum settings to be calculated. The program interfaces with the start/stop, load shedding, and reset programs to modify operation of equipment. Optimizing programs can provide economizer cycle control of outdoor air according to enthalpy and loads. Chilled water and boiler water temperatures can be modified to provide optimum equipment operation according to load. Some optimizing programs are available from the manufacturer's library but others may have to be custom written. (See Figure 6-7.)

Figure 6-7 Interface Between Computer Programs

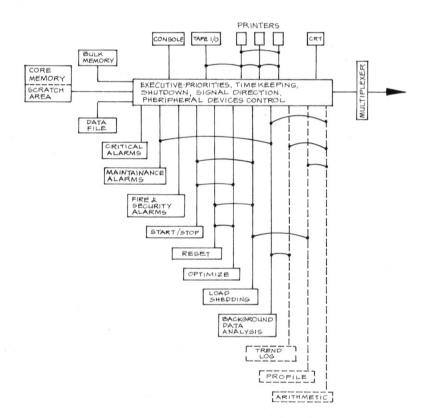

Reset Programs

These programs reset the control points of control thermostats and pressure controllers and interface with the load shedding and optimizing programs. In place of, or in addition to, turning off equipment for load shedding and energy conservation, loads can be trimmed, i.e., if a peak load is approaching, space conditions can be reset to reduce cooling or heating loads; outdoor air dampers also can be reset.

Alarm Programs

These programs report alarms generated either outside the computer (fire alarm pull boxes, boiler high pressure, etc.) or in computer software. Upper and lower limits can be set in software for values of any monitored point and arranged to report an alarm when exceeded.

Alarm programs also can be interfaced with the start/stop program to monitor hours run and report a maintenance alarm when a designated number of running hours is exceeded. Information on particular maintenance required can be stored in memory and printed out when the alarm is reported.

Chapter 7

Developing Electrical Energy Usage and Patterns

ELECTRICAL ENERGY USAGE ANALYSIS

An analysis of electrical energy usage must be performed throughout an operation so management can focus on critical areas where power consumption can be reduced.

To establish a meaningful program for reduction of energy consumption within a facility, it will be necessary to obtain at least a two-year billing history — preferably five years. Each bill should contain such information as kilowatts consumed per hour (kwh) during a certain period of time, the power factor, the billing period and the number of days in each billing period.

Each year should be plotted on graphs (See Figures 7-1 and 7-2), one graph to illustrate kwh for each month in the year, the other to illustrate the dollars expended for each month. From each billing period, plot the monthly demand consumption and demand charges. From available blueprints, plot the MBTU (thousand BTU) consumption per square foot/year and total cost per square foot/year. To arrive at these data, add the kwh for each month in the year. The product of the total kwh/year multiplied by the BTUs per kwh (3,414) will be the total MBTU/year. Divide the total MBTU/year by the gross area in square feet (gsf), to come up with MBTU/gsf/yr. Add up the cost in each month based on kwh consumed. The total cost/year divided by the gross area in square feet will be the total cost/gsf/yr. A picture now has been developed that reflects consumption, cost, demand, etc., visually. It is extremely important to review these graphs with the local power supplier, noting on the graph any rate changes that have occurred. Note also any additions or deletions of square footage on the graph. At this point, request copies of all weather data for the periods covered in the graphs. This information can be obtained from:

National Oceanic and Atmospheric Administration Center
Asheville, North Carolina 28800

or contact the nearest center that can supply enough data for the specified periods.

Plotting Climatic Conditions

Plot the cooling and heating degree days for each month covered in your audit. (See Figures 7-3, 7-4). Compare climatic conditions to energy consumption. While comparing consumption and degree days, it would be well to note wind velocity and solar radiation. This now provides a complete picture of energy consumption and how climatic conditions may affect it. Establish an energy reduction goal and plot this on a monthly basis. (See Figure 7-5).

The goal may be a 5 percent, 10 percent, or 20 percent reduction of energy; plot that goal out and measure your successes month by month.

In reviewing each monthly billing period, verify that all meters are included in each billing. Talk with the local power supplier to determine if a consolidation of meters will effect a cost savings. This must be reviewed carefully, because a consolidation can increase demand, which could well have an offsetting disadvantage in the reduction of electrical energy consumption.

Figure 7-1 Energy Consumption Profile

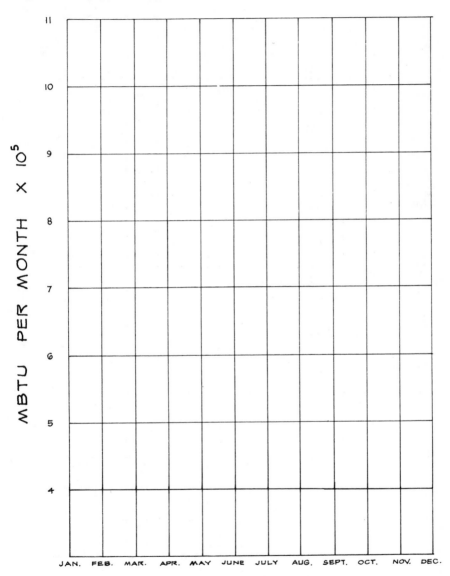

Figure 7-2 Energy Cost Profile

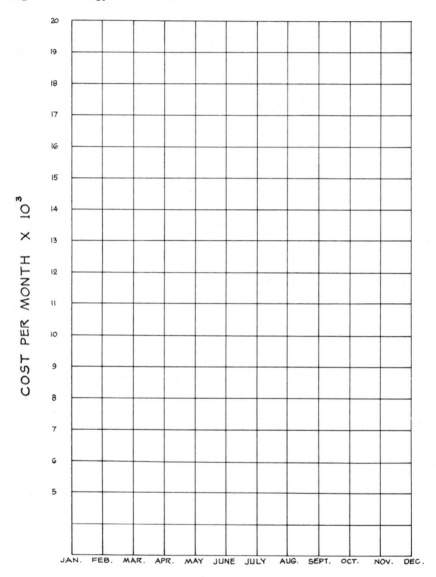

Figure 7-3 Cooling Degree — Days

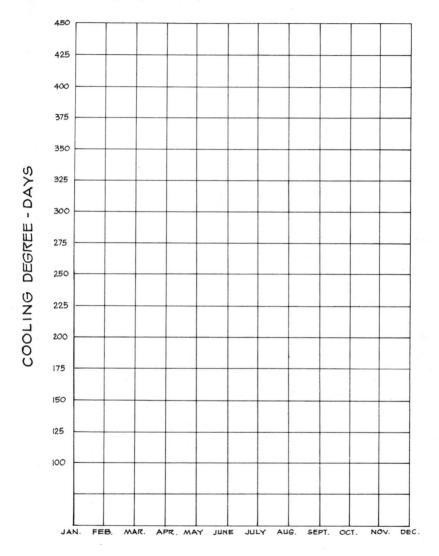

Figure 7-4 Heating Degree — Days

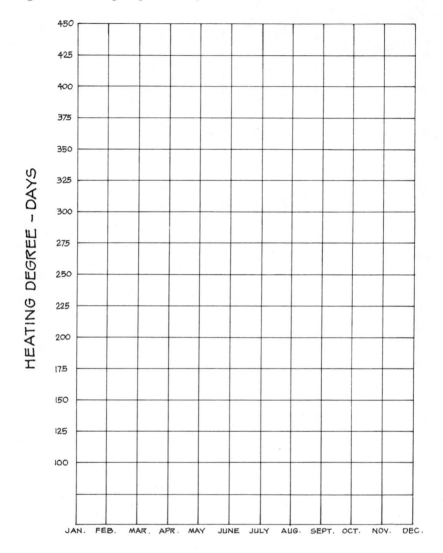

Figure 7-5 Energy Reduction Goal

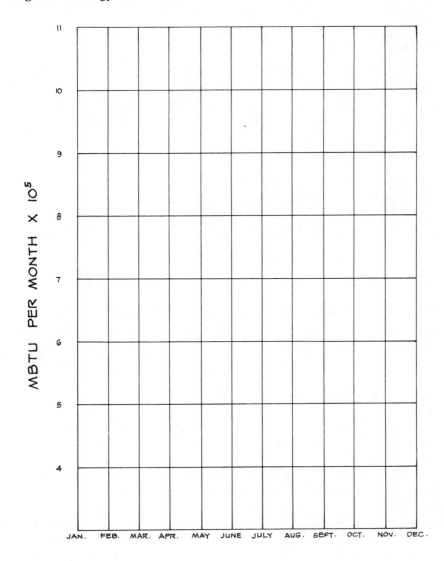

Later, as demand limiting is covered, this must be reviewed and cost estimated to determine the net savings as a result of consolidation.

POWER FACTOR

Power factor is defined as the projection between real or measured watts (also known as effective power) and the volt-amperes (also known as "apparent watts"). If the power factor (PF) is not indicated on the company's billing, check with the local power supplier to determine the billed power factor.

Watts

Power factor (PF) — volt × amp. or
Power factor (PF) — Cos (tan 1 Q) Q — reactive power
(P) P — active power
To measure PF you need only a voltmeter, ammeter, and wattmeter. Example:

$$\text{Power factor} = \frac{345}{5 \times 115} \text{ or } \frac{345}{575} = 60\%$$

In the case of a three phase (∅) 230 volt motor consuming 12 amperes, the volt-amperes are 1.73 × 230 × 12 or 4,775. If the wattage as indicated by the wattmeter is 3,950:

$$\text{Power factor} = \frac{3,950}{4,775} \text{ or } 82.3\%$$

The power factor of a motor improves (increases in percentage) with the rise in horsepower (HP) of the motor and also varies considerably with the type and quality of motor.

It is most desirable to have a high power factor since power suppliers, when furnishing power to establishments where the power factor is low, charge not only for the kilowatt hours consumed, but also make an extra charge based on the kilovolt-amperes used during the month. It is advisable to maintain a PF of 95 to 99 percent. The theory of PF correction is beyond the scope of this work; however, actual correction is accomplished by means of capacitors and/or synchronous motors. The required calculations can be difficult, so it is recommended that a qualified electrical contractor or consulting engineer be called in if the capabilities do not exist in the plant. Low power factor can increase costs by approximately 10 percent. PF correction can pay for itself within a short time.

Carefully examine all consumption and demand factors to determine any abnormalities. If this investigation raises questions, ask the local power supplier for satisfactory answers and in some cases substantial rebates. Always review charges carefully. It is not abnormal to find cases where bills have been paid twice.

With the information gained thus far, prepare a written report for yourself outlining PF corrections, estimated costs versus savings, a review of consolidation of meters weighing cost versus demand increase (if any) versus savings obtained. The graphs should establish trend lines that are occurring in your operation.

Optimizing Power Factor

Excessive voltage drop from the incoming power source at the transformer to the point of use can be a waste of energy. Excessive voltage drop can be caused by several factors:

- undersized wiring
- high resistance
- excessive current

If the power factor of an industrial plant electric distribution system is held at 100 percent at all times, then all of the power transmitted is used. Few industrial plants operate near 100 percent power factor.

Plants where production processes require a high percentage of electrical resistance load can operate with a very satisfactory power factor. Induction motors usually are the greatest consumers of plant electrical energy and result in a high consumption of inductive power. This results in a plant's power factor much lower than 100 percent. It is not unusual to find power factors as low as 60 to 70 percent.

At horsepowers of 200 and above, synchronous rather than induction motors should be installed. Synchronous motors provide leading reactive kilovolt-amperes when operating with full field excitation that offsets the logging kilowatt-amperes inherent with low power factor. Installation of static capacitors to offset the logging kilovolt-amperes directly with inductive electric equipment is another conservation approach.

Static capacitors have the advantage of availability in a number of standard kilovolt-ampere ratings. The devices have no moving parts and can be installed in the plant's distribution system or directly in connection with a major potential power loss source.

DEMAND LIMITING

The demand factor or demand is the ratio of maximum electrical demand to the total connected load. This will depend, of course, on the type of building and its use. The demand factor for any particular class of installation decreases as the floor area increases.

Depending on geographical location, the kilowatt demand charge will vary; however, in all cases the charge per kilowatt of demand is exceedingly high. Because of the high costs involved, demand control techniques are required.

Load factor, the ratio of energy use (kwh) to highest demand (kw) × time, commonly 720 hours (the number of hours in 30 days) measures the effectiveness with which energy is used.

If, for example, 500,000 kwh were consumed in a 30-day period, and during that time the peak demand was 875 kw, a load factor of 79.4% would be established.

$$\text{Load factor} = \frac{\text{energy use (kwh)}}{\text{highest demand (kw)} \times \text{time (hours)}}$$

$$\frac{500,000 \text{ kwh}}{875 \times 720 \text{ hours}} = 0.7936 - 79.4\%$$

The lower the demand, the higher the load factor and, therefore, the relative cost for electrical energy. Demand limiting is a technique used most frequently to improve load factor. This refers to the electromechanical process of load shedding.

To better understand the potential for improving load factor, a complete analysis of demand records indicating hours of peak demand is in order. It is necessary to monitor demand, which can be accomplished either by leasing or purchasing a demand recording meter. In most cases a meter can be obtained from the local utility or electrical supplier at a nominal rental fee. Demand peaks can be identified easily as shown in the following illustration: (See Figure 7-6.)

To control load factor, thereby controlling demand, varied data are required. To develop the necessary information, the rating of each electrical load is required. If complete electrical drawings are not available, a power survey is in order. In evaluating the electrical load, it will be necessary to distinguish between the primary and secondary loads. The primary load cannot be interrupted but all secondary loads can. Some examples of secondary loads are:

- air handling units
- cooling units
- heating units
- machinery
- electric boilers
- pumps
- supply fans
- exhaust fans, etc.

Figure 7-6 Demand Control Peaks

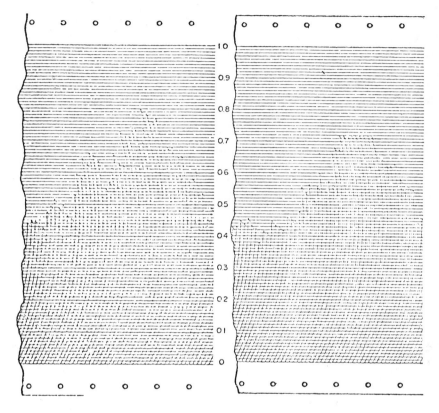

Demand control (without on left, with on right)

After developing the secondary loads, it would be advisable to assign them priorities in terms of first off, last on. Areas must be reviewed to determine the most feasible plan for shutdowns of short durations.

There are several ways to control demand, from simple timers to sophisticated logger control centers. After determining demand profile and load shedding schedules, it is a good idea to examine the economic feasibility of various control types.

Where feasible, install timers to shut down the system according to a predetermined schedule. Areas not used during certain periods of the day or weekends, including both supply and exhaust units, should be shut down entirely. These areas can be sequenced to come on one or two hours before occupancy. In areas where equipment is controlled, it is important to stagger the units at preselected intervals to avoid excessive electrical loading of the system.

For systems not in operation, the savings can be calculated as follows:

$$\frac{(\text{Volts})\ (\text{amps})\ (1.73)\ (\text{hours not operating})\ (\text{rate})}{1,000} = \text{Daily savings}$$

(Daily savings) (365) = Annual savings

As an example, it would be feasible to shut down air-conditioning compressors for a short time. This would allow restarting without any chance of losing the building load. The entire building must be evaluated to determine all areas that can reduce consumption for short periods. If demand charges are excessive, it would be worthwhile to look into the possibility of demand limiters. This kind of decision must be weighed carefully to ascertain that the maximum benefit is derived and the cost/benefit ratio is in your favor. As a good rule of thumb, any equipment that can pay for itself within five years probably is worth acquiring. In evaluating these decisions, it may be advisable to obtain the advice of competent consultants.

Logic Systems

Logic refers to the operating method by which an automated demand control system measures energy consumed and limits demands by shedding certain secondary loads.

Automated demand control equipment ranges from low-cost minicomputers to the highly sophisticated large-scale computers. Where the most serious approach to control demand is taken, an automated demand control computer system is the most effective.

The five logic systems generally used are:

1. ideal rate
2. continuous integral
3. preductive
4. converging rate
5. instantaneous

ELECTRICAL DISTRIBUTION

The choice of power distribution system is determined by the type of power available and by the nature of the load. To transmit power over a given distance with a specified power loss — I^2R (current squared/resistance) — the weight of the conductor varies inversely as the square of the voltage. For example, incandescent lamps will not operate economically at much higher than 120 volts; the most suitable voltages for direct current (DC) motors are 230 and 550 volts, although

550 volts is practically obsolete except for railway motors. For alternating current (AC) motors, standard voltages are 220, 440, and 550, three-phase. When power for lighting is distributed in a district where the consumers are relatively far apart, alternating current is used. The current is distributed at high voltages (1,150, 2,300, 4,000, 6,900 or 13,800) and transformed at the consumer's premises or converted by transformers on poles, in manholes, or in vaults under the street or sidewalk, to 230-115 volts three-wire for lighting and domestic customers and to 230, 440 and 550 volts, three-phase, for power. In thickly settled city districts where large cables are necessary, direct current usually is distributed from substations.

Direct current is best suited for elevator and printing press motors that constitute a considerable portion of the power load. Alternating current produces considerable reactive voltage drop in the cables, which gives poorer voltage regulation. Direct current systems were installed universally during the time when a storage battery reserve was considered necessary. With two or more sources of supply and with highly improved protection from short circuits by quick-acting relays, AC systems have become highly reliable and, wherever economically feasible, are replacing DC systems.

Series Circuits

Where the devices to be supplied with power are of nearly the same current rating, located relatively far apart, and used simultaneously, it often is more economical to supply power at constant current than at constant potential. To operate at constant current, the power-consuming devices and the power source are in series with the same current. In cutting any individual device out of service, an equivalent resistance must be inserted in its place to maintain the same current in the circuit, or some means must be provided at the power source, to adjust the total voltage automatically so as to maintain constant current.

If in a series circuit operating with direct current the resistance of each of the power-consuming devices is R_d the resistance of the line is R, the current is I, the generator voltage when all the devices are operating is E, and the number of the receiving devices is n, $E - nIR_d + IR$, where $IR = e$ is the resistance drop in the line. Now $e = pIl/A$, where p is the resistivity of the material of ohms per cir-mil-foot, l the length of the conductor in feet and A the sectional area of the conductor in cir mils. If the permissible voltage drop e has been determined, the proper cross section of conductor $A = pIl/e$. For mechanical reasons, conductors smaller than No. 6 American Wire Gauge (AWG) are not used generally.

This method of distribution is used exclusively for streetlighting systems in which the lamps are located over considerable distances. The power is supplied by a transformer that maintains constant current irrespective of the number of lamps in the circuit. The necessarily high voltage is not objectionable as circuits placed on poles or underground can be operated safely and satisfactorily at 1,100 to 3,000 volts per circuit. Standard currents are 4.4, 5.5, and 6.6 amps. In some systems it is more economical to supply streetlighting power from the low-voltage parallel system, time switches being used to connect and disconnect the lamps.

Parallel Circuits

Power usually is distributed at constant potential and all the devices or receivers in the circuit are connected in parallel, giving a constant potential system.

If conductors of constant cross section are used and all the loads (L_1, L_2, etc.) are operating, there will be a greater voltage IR drop per unit length of wire in the portion of the circuit AB and CD than in the other portions. The voltage will not be the same for the different lamps but will decrease along the mains in accordance with the distance from the generating end. (See Figure 7-7.)

Figure 7-7 Parallel Circuit

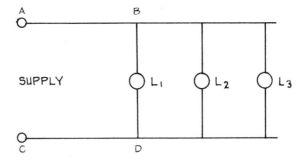

Loop Circuits

A more nearly equal voltage for each load is obtained in the loop system. (See Figure 7-8.)

The electrical distance from one generator terminal to the other through any receiver is the same as that through any other receiver, and the voltage at the receivers may be maintained more nearly equal but at the expense of additional conductor material.

Figure 7-8 Loop Circuit

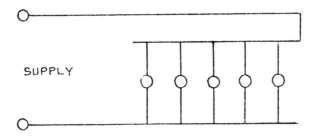

Series-Parallel Circuits

For incandescent lamps the power must be at low voltage (115 volts) and the voltage variations must be small. If the transmission distance is considerable or the loads are large, a large or perhaps prohibitive investment in conductor material would be necessary. In some special cases, lamps may be operated in groups of two in series. (Figure 7-9.)

Figure 7-9 Series-Parallel System

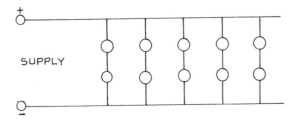

The transmitting voltage is thus doubled and, for a given number of lamps, the current is halved, the permissible voltage drop (IR) in conductors doubled, and the conductor resistance quadrupled. The weight of conductor material thus is reduced to 25 percent of that necessary for simple parallel operation.

Three-Wire System

In the series-parallel system, the loads must be used in pairs and both units of the pair must have the same power rating. To overcome these objections and at the same time to obtain the economy in conductor material of operating at higher voltage, the three-wire system is used. It is devised by adding a third or neutral wire to the system shown in Figure 7-9. The result is illustrated in Figure 7-10.

If the neutral wire is of the same cross section as the outer two wires, the system requires only 37.5 percent the copper required by an equivalent two-wire system. Since the neutral ordinarily carries less current than the outers, it usually is smaller and the ratio of copper to that of the two-wire system is even less than 37.5 percent.

Figure 7-10 Three-Wire System

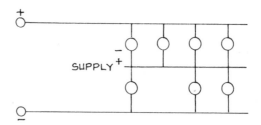

When the loads on each half of the system are equal, there will be no current in the middle or neutral wire, and the condition is the same as that shown in Figure 7-9. When the loads on the two sides are unequal, there will be a current in the neutral wire equal to the difference of the currents in the outside wires. For example, if each of the loads in the system shown in Figure 7-11 takes 1 amp, the current in each part of the system will be given by the numbers on the ammeters shown connected in the system.

Figure 7-11 Three-Wire System With Two Generators

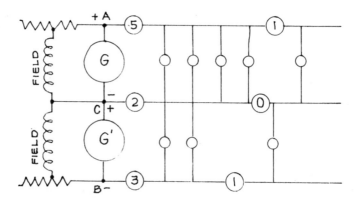

The three-wire system shown in Figure 7-10 is not practicable because no means are provided for holding the neutral at its correct potential. If, for example, four loads are in operation on one side of the system and three on the other, as shown, the voltages on the two sides of the system become seriously unbalanced and the three loads, which may be lamps, are subjected to overvoltage. One method of supplying the neutral in a DC system is shown in Figure 7-11, where each side of the system is supplied by a separate generator. This is open to the objection of the greater complications of two machines, greater cost, more floor space, and the lesser efficiency of two machines.

Balancer Set

Another method of obtaining the neutral is to use a balancer set. This consists of two smaller shunt or compound machines coupled together with the armatures connected in series across the outer lines (Figure 7-12). When the loads are balanced, there is no neutral current and the two machines merely run idle as motors, being connected in series across the outer conductors. If the load on one side of the system becomes greater than that on the other side, the machine on the heavier load side operates as a generator and pumps some of the neutral current to its side of the line. The remainder of the neutral current goes through the other machine, supplying it with the power that enables it to operate as a motor and drive the generator. For example, in Figure 7-12, the load on the positive side of the system is greater than that on the negative side. Hence, machine G on the positive side is operating as a generator and machine M is operating as a motor.

Figure 7-12 Motor-Generator Balancer

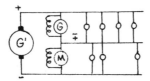

If the machines are compounded so that when operating as a generator the winding is cumulative (or differential when operating as a motor), the voltage imbalance with change in load can be made practically zero.

Since balancer sets take power continuously, they are used for the most part on large systems of high diversity where the percentage imbalance is small, so that the power rating of the set is but a small percentage of the power of the system.

Figure 7-13 Three-Wire Generator

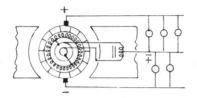

Three-Wire Generator

The three-wire generator is the most common and efficient method of obtaining a neutral in a DC system. It is a conventional generator that ordinarily would be used to supply the outer conductors with power. However, two or more taps, "a" and "b", as shown in Figure 7-14 are brought out from the armature winding to two slip rings (Figure 7-13).

A compensator or reactance coil of low resistance called a balance coil is connected across the slip rings. The neutral of the three-wire system is connected to its center point. The voltage across the slip rings is alternating. Because of its choking action, little alternating current flows in the balance coil. The unbalanced direct current in the neutral flows back through the balance coil to the armature. The inductance of the balance coil has no effect on the steady direct current and the resistance of the balance coil is low, so there is little voltage drop due to the direct current. Two or more balance coils with their neutrals connected may be used. With two such balance coils, the second balance coil is connected to slip rings that are tapped to the armature winding at points 90 electrical degrees from the first.

Feeders and Mains

Where DC power is supplied to a large district, improved voltage regulation is obtained by having centers of distribution. Power is supplied from the station bus bars directly to the centers of distribution by large cables known as feeders. Power is distributed from the distribution centers to consumers through the mains. As there are no loads connected to the feeders between the generating station and the centers of distribution, the voltage at the latter points may be maintained constant. Pilot wires from the centers of distribution often run back to the station voltmeter, assisting the operator in maintaining the potential constant at the centers. This system provides means for maintaining close voltage regulation at the consumer's premises.

Alternating Current Three-Wire Distribution

With the exception of the thickly settled districts of a few of the large cities, where power still is distributed as direct current, practically all energy for lighting and small motor work is transmitted at 1,150, 2,300, or 4,000 volts AC to transformers that step down the voltage to 230 and 115 volts for three-wire domestic and lighting systems as well as 230, 440, and 550 volts, three-phase, for power. For the three-wire systems, the transformers are designed so that the secondary or low-voltage winding will deliver power at 230 volts, and the middle or neutral wire is obtained by connecting to the center of the midpoint of this winding (Figure 7-14).

Figure 7-14 Three-Wire 230-115 Volt AC System

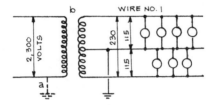

Figure 7-15 208-120 Volt Secondary Network (Single Unit)

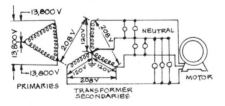

Grounding

The neutral wire of the secondary circuit of the transformer should be grounded on the pole (or in the manhole) and at the service switch in the building supplied. If as a result of a lightning stroke or a fault in the transformer insulation the transformer primary circuit should become grounded (as in Figure 7-14) and the transformer insulation between primary and secondary windings broken down at "b", and if there were no permanent ground connection in the secondary neutral wire, the potential of wire No. 1 would be raised 2,300 volts above ground potential. This would constitute a very serious hazard to life for persons coming in contact with the 115-volt system. The National Electrical Code requires the use of a ground wire not smaller than No. 6 AWG copper. On secondary circuits, grounds should be provided at least every 500 feet. With the neutral grounded (Figure 7-14), voltages to ground on the secondary system cannot exceed 115 volts.

A common and economical method of supplying business and thickly settled districts with high load densities is to employ a 208-120 volt, three-phase, four-wire low voltage AC network. The network operates with 208 volts between outer wires giving 120 volts to neutral (Figure 7-15). Motors are connected across the three outer wires operating 208 volts, three-phase. Lamp loads are connected between outer wires and the grounded neutral. The network is supplied directly from 13,800-volt feeders by 13,800-208-volt three-phase transformer units, usually located in manholes, vaults, or outdoor enclosures. This system thus eliminates the necessity for transformation in the substation. Many such units feed the network so that all the secondaries are in parallel. Each transformer is provided with an overload reverse-energy circuit breaker (network protector) so that a feeder and its transformer are isolated if trouble develops in either. This system is flexible since units can be added or removed easily in accordance with the rapid changes in local loads that occur, particularly in downtown business districts.

Voltage Drops

In AC distribution systems, the voltage drop from transformer to consumer in lighting mains should not exceed 2 percent in first-class systems, so that all the lamps along the mains may operate at nearly the same voltage and the annoying flicker of lamps will not occur with the switching of appliances. This may require a much larger conductor than the most economical size. In transmission lines and in feeders where there are no intermiedate loads and where means of regulating the voltage are provided, the drop is not limited to the low values that are necessary with mains, so the matter of economy may be given consideration.

Maintenance Recommendations

All electrical motor control centers should be inspected and cleaned annually. Discolored terminals are indications of overheating. Starter contacts should be checked for signs of arcing. All terminals should be cleaned, and contacts requiring replacement should be replaced. All motors should be checked with a megger meter and readings recorded. Check with the manufacturer for recommended megger levels. Higher than usual readings indicate that the windings are dirty and/or need replacement. This can be accomplished by a local motor repair service. As the megger readings increase above normal, energy consumption increases proportionally. Check the accuracy of the main facility metering system. Deenergize transformers that could be off during certain periods.

LIGHTING

It is important to optimize the plant's lighting system. Light abatement can be included in the design to minimize energy consumption by specifying the correct type of lighting, such as:

- pendant
- hanging reflective and flood lighting fixtures
- mercury-vapor type
- sodium-vapor type

Whatever the type; it should be color-corrected as necessary. The lighting system design may specify newer types of fixtures that will give the same amount of illumination as incandescent lighting with less energy consumption.

Lighting is of prime importance in attaining the goal of conserving electrical energy. First, obtain all available "as built" information about the lighting system specifications, reflected ceiling plans, floor plans, results from previous lighting surveys, etc., to assist in the lighting analysis.

Existing conditions and the interrelationship between systems must be examined carefully. Office buildings and typical office areas should be visited and light meter readings should be made at the working level of all working spaces. Record all gathered data on the working drawings. Note specific fixtures in use, the number of lamps, supplementary room lighting, etc., that may affect lighting calculations as to quality and intensity as well as to energy usage. Note, too, the colors and reflectance value of wall coverings, floors, and ceilings. These data should be included on the drawing.

By maintaining a higher light output than required it may be possible to reduce wattage of lamps to each fixture, and in some cases the number of fixtures in service, without a reduction in allowable illumination. When relamping, it is desirable to replace existing units with types that use a lower lamp lumen depreciation. Fluorescent lamps with better lamp lumen depreciation (LLD) factors may cost more per unit, but the investment can be recovered by lower expenditures for energy and maintenance.

Survey the building and list all room areas and number of occupants and their activities in each space and the length of time each space is occupied. A careful analysis will reveal many opportunities for reducing the intensity of illumination or shutting off lights completely. Install key switches where opportunities exist for their effective use. When new building additions or extensive remodeling are undertaken, provide a sufficient number of switches so that all lights not needed for periods of time during the day can be turned off. Existing windows can be used advantageously for daylight and thus reduce electrical lighting requirements.

Review existing lighting system, taking into account individual fixtures that can be fully disconnected or lamps that can be removed to reduce wattage of an individual fixture and/or polarized lenses that can be used to improve lighting quality while reducing electrical consumption. Higher reflectance colors also are effective in reducing electrical energy usage. The lighter the color of the room finishes and furnishings, the lower the light absorption by these objects and hence fewer watts per square foot will be required to produce the same footcandles. The use of phantom tubes also is recommended for hallway and storage areas.

An example of a lighting analysis to calculate the percentage of total energy saved in an area totaling 150,000 square feet is shown in Table 7-1.

Table 7-1 Lighting Analysis for Calculating Energy Saving

Location	Existing lighting system 1	Replacement lighting system 2
Building area	10% fluorescent	90% fluorescent
Lighting	90% incandescent	10% incandescent

Consumption	Existing lighting system 1.	Replacement lighting system 2.
Avg. kwh/month	9,790 kwh/month	5,875 kwh/month
Yearly consumption	117,480 kwh/year	70,500 kwh/year

Total energy saved/year (kwh) 46,980
Percent savings/year 39.9%
1. Existing lighting consumption in kwh/year multiplied by the present rate/kwh = current operating cost
2. New replacement lighting consumption in kwh/year multiplied by the present rate/kwh = new operating cost = savings

Ballast Operation Lamps Removed

The energy crisis has caused users to look for ways to reduce power consumption. One method is to remove lamps from energized fixtures. When both lamps are removed from a two-lamp ballast circuit, a small amount of power still is consumed as the ballast draws magnetizing current. Table 7-2 shows this amount:

Table 7-2 Watts Loss for Major Ballast Lines

Two-Lamp Family	Watts Loss Per Ballast with Both Lamps Removed
430 milliampere rapid start . . . . .65.0 w	
800 milliampere high output .12.5 w	
1,500 milliampere power groove . . . . .13.5 w	
425 milliampere instant start (single plan) . . . . 0.0 w	

Source: General Electric Company

System Performance

Power factor will remain above 95 percent provided not more than half of the lamps in an installation are removed.

Ballast life is not adversely affected if both lamps are removed. However, removal of only one lamp from a two-ballast circuit (except instant start) can affect ballast life seriously and can cause immediate (hours) failure.

It is important to note that most manufacturers do not recommend removal of one lamp for any reason. This is an abnormal condition and voids the product warranty.

Lamps from single lamp fluorescent, mercury, and metal halide ballasts may be removed with no adverse effect. Do not remove high pressure sodium lamps or the starting circuit will be damaged. If the lights cannot be switched off, the ballast should be disconnected from the line.

Reduction of operating hours wherever feasible is being encouraged. In some areas where reliable manual switching cannot be effective, automatic timers or photo cells should be provided. These energy savers are essential for parking areas, industrial yards, school protective lightings, etc. Installation of control site lighting and parking lot lighting also is recommended. Table 7-3 shows the recommended lighting levels for indoor work areas.

In conjunction with previous usage history, insight has been gathered based on calculations as to how the new lighting usage patterns affect overall electrical demands and charges and what the total cost savings will be.

Table 7-3 Recommended Maximum Lighting Levels

Task of Area	Foot-Candle Levels	How Measured
Hallways or corridors ....10 ± 5		Measured average minimum 1 foot-candle
Work and calculations areas surrounding work stations ....30 ± 5		Measured average
Normal office work ....50 ± 10		Measured at work station
Prolonged office work (somewhat difficult))....75 ± 15 Prolonged office work (visually difficult)		Measured at work station
Industrial tasks ....100 ± 20		As maximum

Source: Illuminating Engineering Society

Chapter 8

Refrigeration Cycle Efficiency

REFRIGERATION EQUIPMENT — A MAJOR ENERGY USER

The requirements of a facility's electrical and mechanical systems account for all energy utilized by that facility. Surprisingly, one part of the complex — the refrigeration system — can result in energy consumption from 30 percent of a utility budget to upward of 70 percent, depending on the application.

Industry-wide, there are approximately 23 million tons of centrifugal and absorption refrigeration machines. Of this total, centrifugal machinery accounts for approximately 80 percent, with absorption providing the balance.

Domestic centrifugal installed tonnage represents approximately 15×10^6 kw. How much of this energy consumption is it possible to save? The answer is around 10 percent.

Low operating costs or operating efficiency do not relate entirely to power consumption. Operating efficiency should include overall maintenance costs.

To optimize efficiencies is to reduce power consumption. To achieve this effect, it is important to be aware of the critical balances that depend on the refrigerant, water, air, and oil. Any of these, out of balance, can have a dramatic effect on operating costs and efficiencies.

Selecting the Best Refrigerant

The proper refrigerant for a given machine takes into account a number of considerations, the most important being the boiling point of the refrigerant and the operating pressures.

The pressure differential between the inlet and discharge of the refrigerant compressor can be thought of as being caused by a column of vapor that, due to its weight, would produce the same pressure differential between the top and bottom of the column. The height of this column, which is called "head," is measured in feet. The lower the head can be made, the lower the power requirements. (See Figure 8-1.)

Figure 8-1 Normal Head Across a Compressor

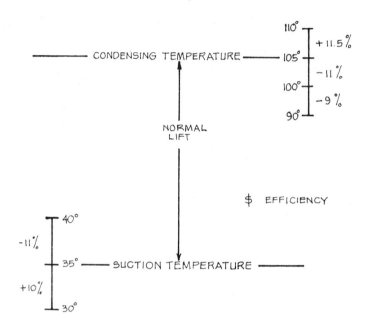

Source: Courtesy of Carrier Air-Conditioning Company

Assume the evaporator pressure is at 35° F and condenser at 105° F. The power consumption can be reduced by increasing the evaporator pressure and by lowering the condenser pressure. Controlling these two pressures at the minimum practical levels achieves the lowest operating cost. Establishing the minimum involves consideration of the air conditioning system used in the space, humidity levels desired, and the machine operating characteristics.

Operating pressures are affected by the conditions that exist in the heat exchangers (Figure 8-2).

Figures such as these are used to depict heat exchanger performances. In Figure 8-2, cool brine enters at condition T1 and exits at T2. The difference between T1 and T2 represents the temperature use across the heat exchanger. The refrigerant temperature or pressure is represented by line TR. The difference between T2 and TR is the leaving temperature difference and is affected by water velocity, water temperatures, and fouling. Figure 8-3 illustrates the conditions that exist in the condenser circuit.

Figure 8-2 Heat Exchanger Pressures

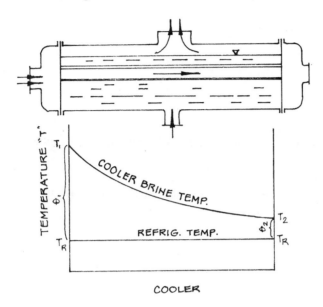

Figure 8-3 Condenser Circuit Operation

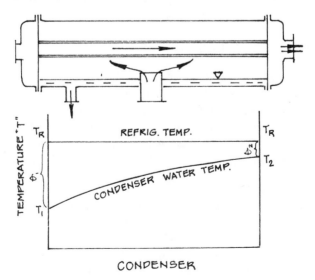

Source: Courtesy of Carrier Air-Conditioning Co.

Figure 8-4   Cross-Section of Tube Depicts Fouling

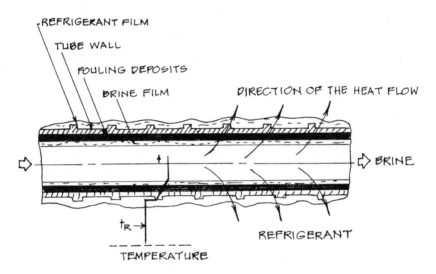

If the leaving temperature is higher than normal, either in the evaporator or condenser or both, the head requirements are increased across the compressor. Once the normal leaving temperature for a given heat exchanger is known, increases in leaving temperatures can be a useful tool in determining if temperature and pressure levels are in balance.

Fouling and Contamination

Figure 8-4 shows the progression of the deposits from a refrigerant film on the outside of the tube to external tube fouling, the resistance of the metal, internal tube wall fouling, and water film coefficient.

The refrigerant film is affected by the refrigerant charge. If the charge is too low, some of the tubes may be moistened inadequately and the leaving temperature difference will be higher than normal. If the refrigerant charge is too high, liquid can be drawn into the compressor before being evaporated. In one case the head requirement is higher and in the other the system would be pumping liquid instead of gas. In either case, the charge is reasonably critical and if not controlled can result in higher than normal horsepower requirements.

External tube fouling can be caused by refrigeratant contamination. If the outside of the tube is fouled with rust deposits that completely fill the fin portion, the result will be extremely poor heat transfer, lower evaporator pressures, and a much higher than normal leaving temperature difference. In addition, high concentrations of oil in the evaporator can affect refrigerion performance and possibly cause external tube fouling.

Water or brine is used as a medium for transferring heat from the air-conditioned space to the refrigeration machine and from the refrigeration machine to the cooling tower. With water, consider fouling, velocity, corrosion problems, and possibly freezeup must be considered.

After the machine has been operating for a time, scale or fouling conditions may form on the inside of the tube walls of the cooler or the condenser. Fouling can be caused by scale, slime, algae, or dirt. Scale is the predominant problem and the internally scaled tube is a typical example. Scale should not be allowed to form, since in addition to increasing power requirements, corrosion or pitting can take place under the scale layer. This type of failure is called concentration cell corrosion. Lack of oxygen under the scale promotes polarization of adjoining surfaces.

Other forms of corrosion, such as dissolution of the metal from straight acidic attack, can occur.

Water treatment, therefore, plays a very important part in overall maintenance practices. The lack of a good water control program can result in fouling, corrosion, excessive maintenance costs, and increased power requirements.

High water velocity causes excessive pressure drop and an increase in pumping horsepower. Too low a velocity can result in poor heat transfer and an excessive pressure difference across the compressor. The importance of velocity cannot be overemphasized since it is a factor in tube corrosion/erosion failures due to unbalanced controls.

Gas bubble entrainment may not be apparent in every case. Normally an inspection once a year on condenser water circuits is practical. Inspection of an evaporator circuit should be performed every five years after the first year of operation.

Another malfunction that must be avoided is freezeup. Insurance loss rates range from 1 to 2½ percent for this type of failure. Tube blockage, valving problems, pump failures, and controls all can contribute to the potential of a freezeup. However, if water flows are maintained, it is almost impossible to freeze a tube.

Water never should be allowed to mix with refrigerant in a refrigeration machine. "Free" water in the evaporator will react with the refrigerant by a process known as hydrolysis to form decomposition products, typically hydrochloric and hydrofluoric acid.

Acidified water provides an electrolyte that links dissimilar metals and can cause galvanic corrosion. Any condition of nonuniformity within the metal, such as may arise from improper annealing or cold working, may increase the heterogeneity and intensity of polarity differences, and the use of dissimilar metals may cause one of them to corrode. Acidified water can cause corrosion and binding of moving parts and severe shell corrosion that breaks off and clogs fins with iron oxide.

Water may enter the machine from several sources, including tube rolls, oil coolers, and air leaks. Keeping in mind the number of machines in domestic operation, it is unrealistic to assume that all joints will remain absolutely tight. In fact, most water found in machines is a result of minute leaks at tube joints, sometimes numbering in the thousands.

Purge units that collect and separate water are installed to give indications of leaks. Purges must be checked periodically to insure that free water is not present. If free water is detected, the source must be determined or operating difficulties will arise. Consideration also should be given to installing desiccants (driers) in the refrigerant circuit to remove free water.

Oils that contribute to maintenance costs are classified into two categories: turbine oils and refrigeration oils.

Turbine oils normally are paraphinic-base crudes. They are not dewaxed. Requirements for dehydration of the oil are minimal. The turbine oils contain rust and oxidation inhibitors as well as extreme pressure additives. These inhibitors are among their most important characteristics since they reduce the possibility of corrosion in the lubrication circuit. Turbine oils have reasonably good foam stability. They are mandatory with machines operating below atmospheric pressure, such as with refrigerant 11 or 113.

In addition to being used in refrigeration machines operating below atmospheric pressure, turbine oils are recommended for gears, both because of their extreme pressure additives and because of the rust and oxidation inhibitors.

Refrigeration oils, on the other hand, normally are napthenic base crudes that are dehydrated and dewaxed and do not include rust and oxidation inhibitors. These oils are used in machines operating above atmospheric pressure where air is not present normally, such as with refrigerant 12, refrigerant 22, and refrigerant 500.

Oil in an operating machine usually is diluted with refrigerant. The possibility exists that the refrigerant either will hydrolize or will break down thermally. When this happens, both organic and inorganic acids will form. The inorganic acids — hydrochloric and hydrofluoric — will cause pitting of the bearing surfaces and general corrosion of the lubricating circuit.

This can be detected by visual examination. Inorganic acids will darken the babbitt metal lining of a bearing and cause rusting in the rest of the lubrication circuit. Laboratory testing of the oil is possible, but it sometimes is difficult to get a representative sample.

One simple test is to keep a sample of the oil that was added initially to the machine. Comparison of this sample visually will give an indication of the condition of the used oil. If samples are submitted for a laboratory examination, a sample of the oil that was charged into the system initially will be available for comparison.

A chemical analysis normally will include a neutralization number that indicates the number of free chlorides present in the oil. This translates into an acid level. An infrared analysis also should be made to determine the type of oil and the contamination level, as well as a spectographic analysis of the ash that will indicate foreign metals.

Equipment manufacturers recommend that oils be changed at least once a year. If an oil with rust and oxidation inhibitors is used, the oil replenishes the inhibitors. In the case of both turbine and refrigeration oils, changing trends to dilute any contamination that is present. If problems exist in the lubrication system, they should be corrected and the oil changed frequently until the problems are resolved.

In systems operating below atmospheric pressure, the leak integrity of the unit must be maintained. As with other problems in the system, air can cause excessive power requirements and contribute to corrosion and subsequent physical damage.

Air raises the total pressure in the system and can indicate fouling. In either case, it does raise the horsepower requirements.

When air is allowed to remain in the system, it contributes to corrosion that results in obstruction of small orifices and passages and adds to fouling on the outside of the heat exchanger tube surfaces. The green appearance of copper oxide is evidence that air is in the system. Here is where the purge unit again comes into play. The purge unit cannot be expected to provide long-term protection against noncondensibles in the system. However, it does serve as an indication that air is present and gives sufficient warning by expelling air, so corrections can be made when the equipment can be shut down.

If the critical balance of refrigerant, water, oil, and air is maintained, it can result in lower power requirements with considerable savings and can prevent a major failure.

The primary goal is to keep the head requirements as low as practicable. By maintaining the low level, the optimum improvement in overall requirements can be realized.

Maintaining the leak integrity of the unit will result in better performance on units operating with low pressure refrigerants and less loss of refrigerant on machines operating with high pressure refrigerants.

Water conditions must be maintained to ensure that fouling does not take place and that the rises in the cooler are as high as is feasible. This ensures that pumping horsepower is kept at minimum levels and that the condenser rise is as low as practicable, with the entering temperature at minimum levels that match a machine's operating characteristics. Water must be kept out of the refrigerant to make certain that internal corrosion does not take place. The compressor oil must be maintained and be of the proper type recommended by the manufacturer. Above all, equipment must be monitored so that one is aware of the conditions that exist and can take remedial action.

Chapter 9

Central Chilled Water Distribution Systems

DESIGNING A CHILLED WATER SYSTEM

A successful central chilled water system saves energy, staff time and mainte-nance costs. To have a reliable pollution-free system, a plant owner should make sure that engineers designing and constructing the system coordinate their efforts. An owner who is not an engineer will need to become familiar with the workings of a chilled water system to assure an effective low-cost operation.

One of the characteristics of a hydronic system (heat transfer by fluids), whether it uses hot, high temperature, or chilled water, is the interdependence of all the elements of the system. The behavior of any single element of hydronic system will affect the performance of the generating plant and distribution system and of all other elements.

In hot water heating systems, incompatible behavior within the operation is not critical since they are limited in size and usually are overdesigned in capacity for the single buildings they serve.

In high temperature water systems, the functioning of every part is critical, but if the parts are not working together smoothly, the operation can be corrected more easily. The high temperature system frequently terminates in heat exchangers that serve low temperature steam or hot water building systems rather than directly into terminal equipment.

In central chilled water systems, compatibility among the parts is essential. There are many such systems in existence that cannot attain design capacity, adequately supply demands of all users, or operate efficiently without wasting energy unnecessarily. The causes of the insufficiency and energy wastage fall into three categories:

1. incompatibilityof design concepts of central building systems because of lack of coordination or lack of understanding of the underlying principles
2. incorrect design of the interface
3. faulty design of the distribution system

105

INCOMPATIBILITY OF CONCEPTS IN DESIGN CAPACITY

Unequal Return Temperatures

The temperature spread between the supply and return equipment designed into a large central chilled water system (usually called the Delta T) has great impact on the total cost. The central system designer seeks to achieve a large Delta T — 14° F or 16° F is common. The building system designer often prefers small Delta Ts to reduce coil sizes and costs because the effects of unbalanced flow are reduced. However, this is a fallacy since the impact of a cost reduction of the coils usually is negative when compared with the resultant cost increase of the overall distribution piping.

While the supply temperature in a secondary system can be made higher than that in the primary through internal blending, the return temperature in the primary can be equal to or lower (but never higher) than the secondary system.

Unbalanced Flow

In constant volume flow systems with a wild coil and no control of flow or in three-way valve controlled coils with bypassing flow, those closest to the central plant often will rob water from users farther away. This also can occur in large variable volume flow systems where commercial type controllers and control valves are installed (often without positioners), as opposed to industrial type control valves. These valves may not be able to hold against the head pressure available in the mains. In both cases there is overcirculation through the first users, causing the Delta T to decrease in the system, and the result is an inadequate chilled water supply at the end of the line.

A popular technique for remedying the imbalance is to install constant flow devices similar to the Griswold valve. However, these devices are effective only in constant volume flow systems. In variable volume flow systems they can act only as a limiting device at maximum load. At partial load, they are totally ineffective.

Incompatibilty in Partial Load Operation

For operating economy, most large systems are designed with multiple refrigeration machines and pumps. The objective of this design is to operate the least number of machines, towers, and pumps as near to the design capacity as required by the air-conditioning demand. This presupposes a variable volume flow in the primary system with the chilled water circulation rate proportional to the refrigeration demand. In other words, the system would maintain a reasonably constant Delta T under all load conditions. It is poor economics and a waste of energy to operate all chillers in a central plant at low capacity, possibly with the hot gas bypass fully open and all pumps and towers running at a constant energy input, simply because the chilled water demand is out of step with air-conditioning demand. Yet this is exactly the case in many existing central systems. They serve constant volume flow secondary systems in which the supply temperature is invariable and the return temperature variable. As noted, the impact on the primary system also is to impose on it a variable return temperature.

Figure 9-1 illustrates the effect such secondary systems have on central plant operation in a simplified example based on actual observation. Assume a central plant of 20,000 tons capacity composed of five machines of 4,000 tons each, in a parallel arrangement, feeding a primary distribution system at 40° F. The building's secondary system is designed for a steady volume flow at a constant supply temperature of 42° F and a return temperature of 52° F at maximum demand.

Figure 9-1 Central Plant Operation

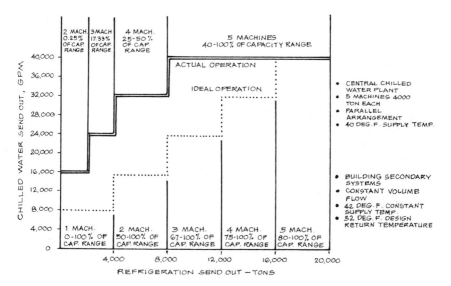

Under partial load conditions, plant operation is as follows:

A minimum of two chillers must be operated when system demand varies between zero and 2,000 tons. The output of each machine is between zero and 25 percent of its capacity. Three machines operating between 17 and 35 percent capacity carry the load between 2,000 and 4,000 tons demand. Four machines take the load between 8,000 and 20,000 tons. This clearly is an inefficient and energy-wasting operation. It is the result of the incompatibility of a constant volume flow secondary system with a variable volume flow primary system.

A drastic remedy would be to control the secondary systems by maintaining the return temperature constant and letting the supply temperature fluctuate. However, this generally is impossible because of the forfeiture of humidity control and other reasons. A sliding supply temperature, such as a control for a constant mean temperature between supply and return, would help somewhat. A better solution would be to change all controls of coil output from three-way bypassing to two-way flow throttling type.

INCORRECT DESIGN OF PRIMARY/SECONDARY
SYSTEM INTERFACE

In many problem systems, the interface design is the cause of the problem. The two examples of primary and secondary systems cited above both require three-way control valves to link the primary and secondary systems together. Most problem systems I have been asked to investigate employ three-way control valves in this interface. Three-way valves generally are not suited to link the primary and secondary systems together.

A three-way valve is a switching device. Whether it is the one-plug mixing or the two-plug diverting type, it cannot function properly unless the fluid pressures in the two branches are approximately equal; in other words, it cannot be used for throttling in one branch. However, in most systems the valve is called upon to throttle against the primary system head pressure.

Under high load conditions the three-way valve will lock the primary and secondary systems into a series pumping arrangement that will upset the flow in the secondary circuit. In extreme cases it has caused the flow in adjacent secondary systems to reverse.

There is only one type of three-way valve interface that functions as intended. This is shown in Figure 9-2.

In this system the supply from the primary system is constant. At partial load the surplus bypasses the secondary system into the primary return. Except in small installations, composed of only one chiller or of two in tandem and one pump, this type is not suitable for central chilled water systems. Although this type of system is not in common use, I have included it to show an application where the pressure in both branches is equal and the valve is able to function as intended.

The following examples illustrate diagrammatically the most common misapplications of three-way valves in the primary/secondary system interface.

Figure 9-3 shows what the designer expected to achieve in a 44° F secondary system connected to a large 40° F secondary system close to the central plant so that the pressure distribution in the mains would not be affected by the behavior of the secondary system.

To obtain a 1,000-ton load, the designer expects to draw 1,500 gpm from the primary system and return it at 56° F. He has sized the pump head of the secondary system circulating pump to match the resistance of the branch piping from the primary system, the secondary system piping, and the pressure drop through secondary system control valves and coils. In other words, the pump is designed not as a circulating pump in the secondary loop but as a booster pump.

Figure 9-2 Three-Way Valve Interface, Proper Function

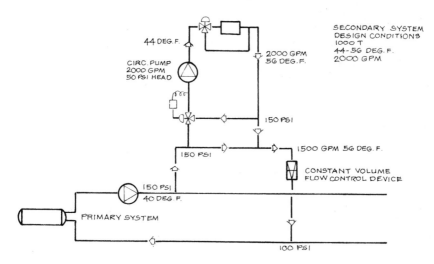

Figure 9-3 Three-Way Valve Interface, Secondary System

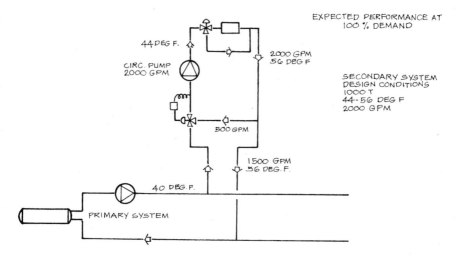

Figure 9-4 Three-Way Valve Interface at 100% Demand

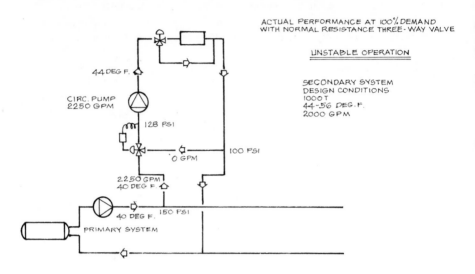

Figure 9-5 Three-Way High Resistance Valve Interface

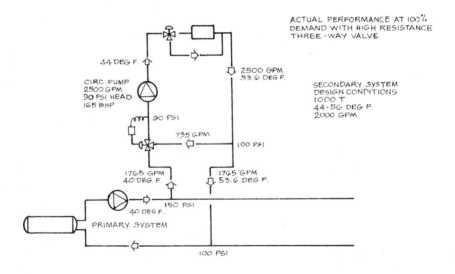

Figure 9-4 shows what actually would happen at 100 percent demand. Since the resistance of the valve is small relative to the primary system pressure head, the result is a negative pressure gradient in the bypass. Therefore, no mixing can take place. The three-way valve starts hunting. The system is unworkable until the valve is blocked in the open position for flow from the primary into the secondary system.

Figure 9-5 shows the behavior of the same system with a high resistance three-way valve. The system will function, but the actual draw on the primary system will increase some 23 percent to about 1,850 gpm and the return temperature will drop to 53° F. In addition, the brake horsepower requirements of the secondary pump increase significantly.

Figure 9-6 Three-Way Valve Interface, Two Secondary Systems

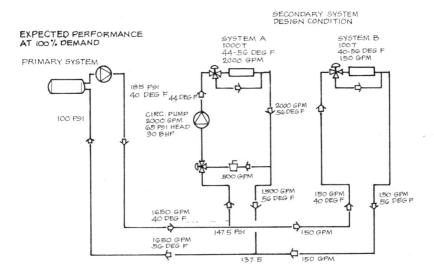

Figure 9-6 shows the designer's expectations from a similar secondary system and an adjacent smaller secondary System B. In this case a normal resistance three-way valve and a resistance in the bypass were provided.

Figure 9-7 Three-Way Valve Interface, Actual Performance

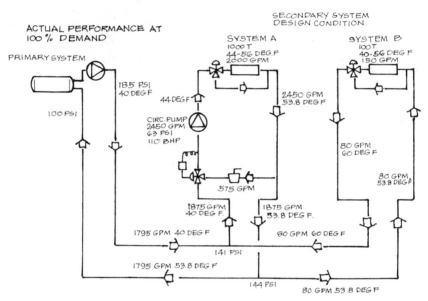

Figure 9-7 indicates the actual behavior of System A and shows that the flow through System B is reversed. In other words, System B does not get any chilled water.

An infinite number of variations of the three-way valve interface can be presented. None of the variations is entirely satisfactory. I have seen few that provide the Delta T design that do not draw more water than the design provided for and none that control the water in a satisfactory manner over the load range. The problem, as mentioned earlier, is that under high load conditions the primary system circulating head, which never is constant, becomes additive to the secondary system pump head, turning the pump into a booster pump. Under low load conditions, when the secondary system circulates within itself, this additional head no longer is available. However, under such conditions, one of the ports of the three-way valve must throttle against a pressure difference, which is the composite of the full primary system pressure head plus the resistance of the three-way valve. In most cases the pressure difference becomes excessive and considerable leakage takes place, further aggravating the poor system performance at partial loads.

In some installations the problems of behavior at full load are overcome by installing a second control valve in the branch piping connecting the primary control over the entire load range. However, as discussed previously, it cannot improve the primary system Delta T as long as the secondary system flow rate is constant and its return temperature variable.

IMPROVED DESIGN OF PRIMARY/SECONDARY SYSTEMS

Figures 9-8 and 9-9 show a design of a secondary system that achieves a reasonably constant Delta T and variable flow in both the primary and secondary systems over the normal operating range by replacing the bypassing control at the coils with throttling control. This system provides the desired compatibility of primary/secondary systems and will permit starting and stopping of machines, towers, and pumps in the central plant in proportion to the load. However, it is not without imperfections.

Figure 9-8 Valveless Interface at 100% Demand

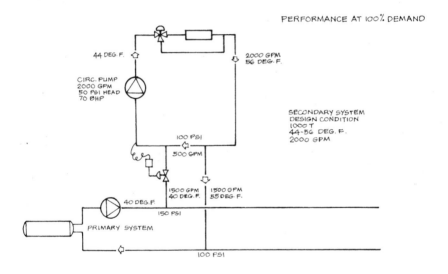

Figure 9-9 Valveless Interface, Alternate 1

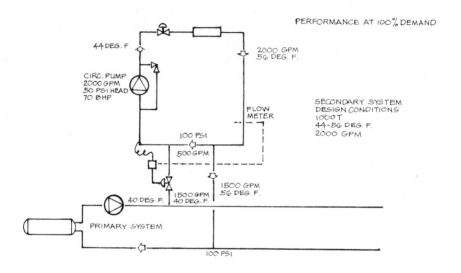

Figure 9-10 Valveless Interface, Secondary System

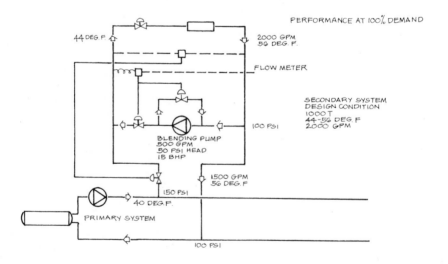

When reducing the chilled water flow through air-conditioning coils in proportion to the load, a transition zone is reached in which the flow pattern turns from turbulent to laminar and the coil performance becomes erratic. For the coils, it is immaterial whether the flow reduction effect is achieved through two-way throttling or three-way bypassing type control valves. Depending on initial velocity, the transition occurs in the 20 to 30 percent range. The phenomenon can be overcome by raising the supply temperature at low loads and letting it float. Since low loads correspond to low flows, a flow meter in the secondary system can be employed to reset the temperature controller at these low loads to maintain a constant minimum flow. Figure 9-9 incorporates the arrangement.

The diagrams indicate that the energy requirements of the circulating pump in the secondary system vary with the resistance at the interface, which was shown to be significantly reduced in the recommended design without a three-way valve interface. Nevertheless, in the interest of conserving energy, the desirability of throttling away the primary system circulating head and subsequent repressurization in the secondary system circulating pump must be considered. Where feasible, the primary system pump head should be used. Figure 9-10 shows the secondary system pump in a temperature control bypass.

The system no longer is a true secondary circuit since both primary system pump and blending bypass pump provide for circulation in it. The control valve in the branch line between the primary and secondary circuits is used now to maintain constant differential pressure at the coils. The controller in the pumped bypass maintains constant inlet temperature to the coils except that, as before, it can be programmed to let the supply temperature float at low loads. The pump is designed for bypassing flow at 100 percent demand. Since at partial load the flow to the coils is reduced, the pump output is lowered accordingly. The brake horsepower of the pump is about on-tenth of that required for equal coil performance in the arrangement shown in Figure 9-6.

CONCLUSION

A good central chilled water system saves energy, personnel time, and maintenance costs, provides reliability, reduces pollution, and allows for greater architectural freedom in building design. Moreover, the design and construction of the system, to be successful, demands know-how, discipline, and the coordinated efforts of the designer of the central plant and primary system, the designers of the secondary systems served by the primary system, and, finally, the owner.

The demands on those involved in creating the system are outlined below:

Central Plant and Primary System Designers

1. thorough acquaintance with the underlying principles of hydronics and their correct application
2. knowledge of actual systems behavior
3. preparation of detailed criteria for the building system designers to follow
4. coordination with the building system designers

Building Systems Designers

1. strict adherence to the design criteria laid down by the central system designers
2. coordination with the central system designers

Owner

1. selection of an engineer for central system design on basis of demonstrated, successful past performance in similar systems
2. enforcement of adherence by the building system designer
3. enforcement of coordination between the two engineers
4. appointment of a knowledgeable engineer who is given the continuing task of updating the master plan and keeping the distribution system model current, and who is empowered to check the building system designs and enforce adherence to the criteria.

The owner cannot supply these capabilities should delegate the functions and the power of enforcement to the central system design engineer.

Chapter 10

Total Management Program Implementation

MONITORING CONSERVATION PROGRAMS

To reduce costs to the minimum, management's energy conservation plan must involve every member of the organization. The most important tasks are monitoring or tracking energy consumption, keeping track of trends, and measuring performance. Goalsetting is useless if the set rate of attaining the goal is not followed.

In accounting, the cost of energy is another important item of information that can be derived from an energy management program. Following the cost trend can be an effective way to control costs and increase profits.

Plotting the rate of energy consumption per unit of production for a portion of a facility or an entire plant is essential in gauging the effectiveness of a program over a period of time. Keeping track of energy usage by itself is not the whole story because it does not reflect the effect of production on energy use. For example, because the energy use over a two-year period might have shown a gradual decrease, a manager might think everything is all right. But if plant output was decreasing more rapidly over this period of time, a comparison with energy usage per unit of output would have shown an increased rate that could be cause for alarm.

Energy usage can be charted in many ways: cubic feet of gas, gallons or barrels of oil, kilowatt hours of electricity, pounds of steam, or a combination of these. If several forms of energy enter the plant, it may be desirable to keep track of total energy consumption over a period. This can best be done by converting cubic feet of gas, gallons of oil, or kilowatt hours of electricity into BTUs (British thermal units), the universally used unit of energy. Then increases or decreases in BTUs can be plotted against time on the plant charts.

Monitoring Examples

Following are examples of methods that can be used for monitoring energy usage. Of course, each organization must evaluate its own needs to determine which method produces the desired results. In fact, several of the methods may be suitable.

Table 10-1 Shows Corporation A's total monthly electricity and gas consumption for 1977

Month	Electricity (kwh)	Gas (million cu. ft.)
January	4,000,000	200
February	3,850,000	180
March	4,100,000	230
April	3,500,000	170
May	3,450,000	170
June	3,200,000	160
July	3,100,000	160
August	3,800,000	180
September	3,050,000	160
October	2,900,000	150
November	2,900,000	160
December	2,900,000	150

Example 1: Corporation Fuel and Electricity Monitoring

Corporation A is a multiplant company with two plants, Plant 1 and Plant 2. In addition, each plant has four separate departments, each manufacturing a specific product. The total electricity and gas consumption for the corporation on a monthly basis for 1977 is shown in Table 10-1. Figure 10-1 charts total electricity consumption by month and Figure 10-2 shows total gas consumption. Similar tables and charts are drawn up for each individual plant and department.

Figure 10-1 Corporation A's Monthly Electricity Consumption, 1977

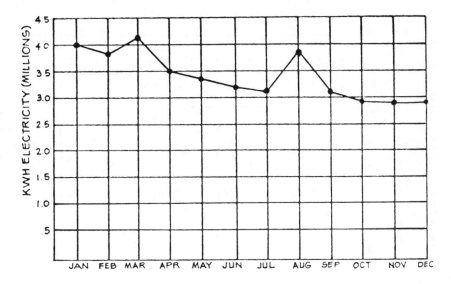

Figure 10-2 Corporation A's Monthly Gas Consumption, 1977

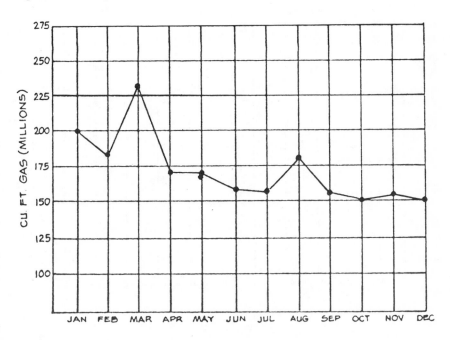

Figure 10-3 Corporation A's Monthly BTU Consumption, 1977

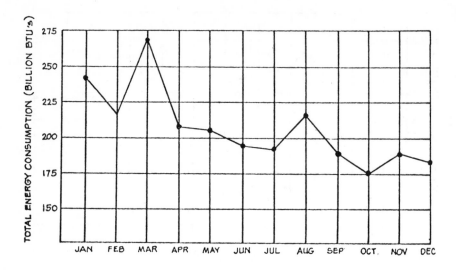

Figure 10-4 Corporation A's Progress in Conserving Energy

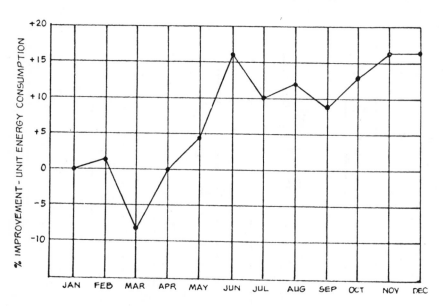

Example 2: Total BTU Monitoring

To track the total energy consumed, the individual energy users must be converted to one common unit, the BTU. For electricity, many plants use the conversion factor of 10,000 BTU/kwh because a power plant takes 10,000 BTUs of fuel to manufacture 1 kwh of electricity. Natural gas varies in BTUs content, but on the average the value is 1,000 BTUs/cu. ft.

Converting the figures in Table 10-1 to BTUs, Table 10-2 shows the total BTUs consumed each month by Corporation A in table form and Figure 10-3 shows the total in chart form. Similar tables and charts are drawn for each plant and department. (Tables 10-3 and 10-4.)

Table 10-2 Corporation A's Total Monthly BTU Consumption (from figures in Table 10-1)

Month	Electricity (billion BTUs)	Gas (billion BTUs)	Total Energy (billion BTUs)
January	40.0	200	240.0
February	38.5	180	218.5
March	41.0	230	271.0
April	35.0	170	205.0
May	34.5	170	204.5
June	32.0	160	192.0
July	31.0	160	191.0
August	30.0	180	218.0
September	30.5	160	190.5
October	27.0	150	175.0
November	29.0	160	189.0
December	29.0	150	179.0

Table 10-3 Corporation A's Monthly Energy Consumption Per Unit of Production

Month	Total energy (billion BTUs)	Total production (1,000 tons)	Total energy per ton (million BTUs/ton—1÷2)
January	240	20.0	12.0
February	218	18.6	11.8
March	271	21.0	13.0
April	205	17.0	12.0
May	204	18.0	11.5
June	192	17.0	11.0
July	191	17.6	10.8
August	218	21.0	10.5
September	190	17.2	10.9
October	175	16.6	10.5
November	189	18.9	10.0
December	179	17.9	10.0

Example 3: Monitoring Total Energy Per Unit Output

Although the information charted thus far is useful to management, it does not tell the entire story of energy consumption. For many industries it is important also to relate BTUs to units of production in order to track energy consumed per unit of production. Table 10-3 shows Corporation A's monthly unit production and energy consumption so as to identify the amount of energy consumed per unit of production. Similar records are plotted for each plant of the corporation and each department.

Example 4: Monitoring Percent of Energy Conserved

It may be useful to keep track of the progress of the conservation program. It is important to establish a base line (a representative monthly energy consumption figure prior to the start of the program). For this calculation, assume that Corporation A began its program in January 1977 and that an average monthly energy consumption figure prior to the start of the program was 12 million BTUs/ton. Table 10-4 shows monthly progress and Figure 10-4 presents the same material in chart form. Similar tables and charts are prepared for each plant and department.

Table 10-4 Corporation A's Monthly Progress in Conserving Energy

Month	Total energy per ton (million BTUs/ton from Table 10-3)		Percent improvement
January	12.0	$\frac{12-12.0}{12} \times 100$	0.0
February	11.8	$\frac{12-11.8}{12} \times 100$	1.7
March	13.0	$\frac{12-13.0}{12} \times 100$	−8.3
April	12.0	$\frac{12-12.0}{12} \times 100$	0.0
May	11.5	$\frac{12-11.5}{12} \times 100$	4.4
June	11.0	$\frac{12-11.0}{12} \times 100$	15.6
July	10.8	$\frac{12-10.8}{12} \times 100$	10.0
August	10.5	$\frac{12-10.5}{12} \times 100$	12.5
September	10.9	$\frac{12-10.9}{12} \times 100$	9.0
October	10.5	$\frac{12-10.5}{12} \times 100$	13.0
November	10.0	$\frac{12-10.0}{12} \times 100$	17.0
December	10.0	$\frac{12-10.0}{12} \times 100$	17.0

Another factor in comparing unit energy consumption with a base energy rate is unit production level. When facilities are not being fully utilized, the level of operating efficiency may not be as high as at full capacity. In other words, at 50 percent capacity, instead of a process' using 50 percent of the energy, it may use 60 percent.

Although one of the key elements of an energy management program is the monitoring of data or the collection of energy cost and consumption information, precise records also must be kept of program developments such as meetings held, projects initiated, and all other activities falling within the scope of the program.

Similarly, the actual quick fix, refit, and systems convert modifications taking place in a program should be monitored. Cost and performance must be controlled. Work must be inspected carefully and occasional checks should be made of operating and maintenance personnel to ensure that their jobs are being carried out as efficiently as possible, especially since an energy management program may require new routines and techniques.

Energy Monitors

Each group of building occupants also can be asked to appoint a person responsible for the facility in the context of the energy management program. The person appointed would help to ensure that building occupants are aware of program objectives and that certain measures are carried out every day. Examples are making sure that lights are turned off during unoccupied hours and that thermostats are set back at night.

PLANT MANAGEMENT CONSERVATION EFFORTS

Management commitment is vital to a successful energy conservation program. This commitment must be active rather than passive, however, and it must be communicated clearly to all organizational levels. To be credible, management commitment must be reflected in words and actions.

One good way to get top management committed to an energy conservation program is to conduct an energy survey in the plant and make energy saving proposals that clearly demonstrate the profitability that can be expected from such actions.

Once top management commitment has been given, these questions must be answered:

1. How is the job of the energy coordinator or energy task force defined?
2. Who should be the energy coordinator?
3. Who should be on the task force?
4. How can energy conservation objectives be communicated effectively to operating personnel?
5. How are operating personnel trained in energy conservation activity that ultimately will meet objectives?
6. Finally, how is the interest of operating personnel in meeting these objectives maintained — in other words, how can they be motivated to continue to conserve energy?

Specific answers to these questions will vary, of course, from company to company. For help in answering them, look at principles of organization and management and the way they relate to the energy conservation process.

One definition of management is "getting things done through other people." The four basic activities of the management process are as applicable to managing an energy conservation program as they are to any other management program. These basic activities are planning, organizing, controlling, and implementing.

PLANNING

Planning has been defined as (1) outlining a course of action to achieve an objective, (2) devising a method to attain a defined goal, and (3) developing a specific way to obtain a desired result. Each definition contains two elements: Knowing what is wanted to be accomplished and deciding how to accomplish it.

The energy planning process consists of three steps:

1. deciding what is wanted to be accomplished, or setting objectives
2. deciding how to accomplish the objectives, or outlining procedures
3. deciding how to apply the skills needed to meet objectives, or assigning responsibilities. This step is necessary because the manager does not carry out the plan singlehandedly. It is the manager's job to get other people to do this.

Management's energy objectives usually focus on the quantity, cost, and effect of energy savings. However, they are not limited to these areas; they will vary by industry, location, and specific situation. Often they will result from changing conditions: a new product may be added, a new plant may be needed, a department may be relocated. All these changes require more energy planning.

To meet change, management makes plans. The first step in the planning process is setting objectives. This also is the first step in the total energy management process. Poor objectives will cause plans to fail. Objectives always should be reviewed to ensure that they are specific, consistent, and attainable.

To be specific, an objective must state clearly what is to be accomplished, when it is to be accomplished, and how it will be measured. Consider the role of the plant manager. The following objectives have been submitted by the department managers. The plant manager must select the objectives felt to be specific.

1. to improve energy conservation
2. to improve boiler efficiency by 2 percent a month for the next four months
3. to reduce steam waste
4. to reduce by January 15 the gas consumption of the process furnace in Unit A (consumer products) by 10 percent

If b and d were selected, the manager saw that they were much more specific because they gave a measure of improvement or reduction.

The second criterion a good objective must meet is consistency. To be consistent, an objective must coincide with the company's overall policies and goals in terms of substance and priority. For example, a plant with a policy of achieving maximum energy savings cannot accept a cost saving objective from the service department manager to skip periodic boiler efficiency checks in order to reduce payroll.

Objectives also must be attainable. If an objective cannot be reached, employees and mangers will become discouraged and the results may be worse than if no objective existed at all. But this does not mean that objectives should be established at a level low enough to guarantee they will be met. Objectives are management's goals and, because they are what management wants to accomplish, they should be realistic. These goals may not be met in every case but everyone must know it is possible to meet them. Review the objectives below and select those that probably are attainable in a well-established machine repair shop:

a. Decrease average energy consumption in Product A by 45 percent within six months.
b. Work overtime during the vacation season to provide repair coverage.
c. Have all new machines checked for safety before they go into operation.
d. Increase energy efficiency 43 percent per hour through closer supervision of personnel.

Of these objectives, b and c probably are obtainable. The second step of the planning process involves outlining procedures. This process forms the framework for how the plan's objectives will be accomplished. Basically, it is a four-step process:

1. defining activities
2. establishing a timetable
3. reviewing resources
4. determining locations

Finally, the third step of the planning process is to assign responsibilities. Who has the skills to carry out each activity? Make as close a match as possible between needed skills and available skills.

ORGANIZING

The next consideration is organizing for energy conservation. Organizing has been described as (1) establishing a structure for the plan, (2) arranging and correlating tasks, and (3) describing a framework of jobs. Organizing often is thought of as dealing with people, but these definitions do not mention people.

They are concerned only with creating a structure of jobs to get the work accomplished as planned. The plan's implementation will have to do with people.

The plan for a successful energy conservation program must be based on several building blocks of organization used in creating job structure:

a. objectives
b. authority
c. responsibility
d. definition

The first building block of organization is objectives. Every job must have a specific objective — a reason for the existence of that particular job. See Table 10-5 for a plant organization chart, complete with the objective for each position given.

Table 10-5 Plant organization chart.

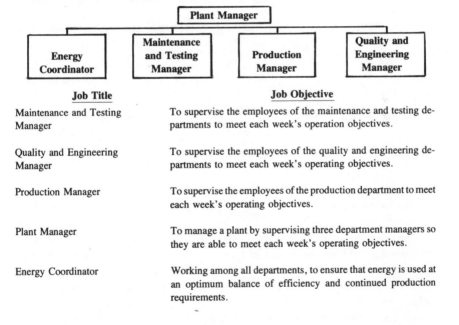

Job Title	Job Objective
Maintenance and Testing Manager	To supervise the employees of the maintenance and testing departments to meet each week's operation objectives.
Quality and Engineering Manager	To supervise the employees of the quality and engineering departments to meet each week's operating objectives.
Production Manager	To supervise the employees of the production department to meet each week's operating objectives.
Plant Manager	To manage a plant by supervising three department managers so they are able to meet each week's operating objectives.
Energy Coordinator	Working among all departments, to ensure that energy is used at an optimum balance of efficiency and continued production requirements.

The second building block of organization is authority, such as the authority that moves downward through the company's chain of command. Managers' authority defines their decision making scope — the kinds of decisions that can be made. Once given authority, the manager is responsible for using that authority to meet objectives.

Responsibility, the third building block of organization, must accompany authority. That is, the manager with the authority to make certain kinds of decisions has the responsibility for the results of these decisions.

Once a particular job or position within the organization has these three building blocks in place, not much has to be done to create the fourth block: definition, meaning that each job must be defined. Whether the definition is called a job description or a position description, it includes statements of the job's objectives, extent of authority, and specific responsibilities. A partial position description for an energy coordinator is shown in Table 10-6.

Table 10-6 Partial position description for an energy coordinator

Building Block	Position Description
Objective	Function: to supervise the activities of the energy conservation group in order to meet each month's operating objectives
Authority	1. to accept or reject applicants from the personnel department 2. to request needed training for employees 3. to order supplies for special equipment 4. to assign work as scheduled and in consonance with the affected department
Responsibility	1. to maintain an adequate number of trained employees 2. to maintain a satisfactory inventory of supplies 3. to meet schedules

CONTROLLING

The third activity of management, controlling, involves supervision — the part of control activity that deals with employee behavior on the job. Earlier, management was defined as "getting things done through other people." Planning involves determining the desired end result, how to accomplish the goal, and which employee(s) will be able to handle the job. Organizing establishes a framework within which to accomplish the goal of the plan. Neither planning nor organizing deals with the behavior of people who operate energy-consuming equipment. However, controllings does. It entails supervision, which deals with ways to improve energy conservation behavior on the part of plant employees. Supervision should be designed to make sure that people produce results that conform to the energy conservation plan.

There are two types of supervisory controls — continuing controls and warning controls:

1. Continuing controls are designed to prevent things from going wrong. A pressure controller, for example, is a continuing control that keeps the pressure in a boiler and steam system at the proper level.
2. Warning controls alert the manager that something may go wrong. A pressure alarm, for example, is a warning control that alerts an operator when pressure is too high.

Two basic concerns of supervision — training and motivation — are continuing types of control. Training, of course, means teaching operators all they need to know to do the job correctly and to conserve as much energy as possible. Motivation means creating job conditions and a job atmosphere that encourage employees to do the job willingly and correctly with a high degree of morale. The well-motivated operator will look for ways to conserve energy instead of wasting it carelessly or maliciously.

A third basic concern of supervision lies in checking on performance. Are energy reduction goals being met? If not, further training or other action may be called for. Clearly, checking on performance is a warning type of control.

Obviously, all employees cannot be supervised on all activities at all times. Managers must spend their checking time in the most efficient and effective manner possible. One helpful technique is periodic inspection. The principle behind inspection is that people do what managers inspect, not what managers expect. When operators notice they are being inspected periodically to see that they are following energy-saving procedures, they know that management means business.

With control by inspection, priority is given to significant exceptions from what is planned. Therefore, thought must be given to controls in management's first activity — planning. Both boundaries and checkpoints are established during the planning activities. Boundaries are the allowable limits within which things can deviate from plan, and checkpoints are the planned times to check progress.

An energy coordinator may plan for a 12 percent energy reduction within six months. However, the coordinator realizes this will be an average figure. In some production units, energy reduction may be greater than 12 percent and in others it may be less. On the basis of experience, the energy coordinator decides that a reduction between 9 and 15 percent will be acceptable. The decision is expressed this way:

Upper limit	15 percent
Objective	12 percent
Lower limit	9 percent

The allowable limits are the boundaries set during the planning activity.

After the boundaries are set, actual energy reduction is tracked. Figure 10-5 shows how one energy coordinator charted actual energy reduction by specific months. This obviously is the result of control by inspection. During these planning activities, the energy coordinator decided to record results on a weekly basis, and the dates became the checkpoints from which the graph was plotted. The energy coordinator would continue to plot energy reduction for subsequent months.

Figure 10-5 Monthly Energy Reduction Within Set Boundaries

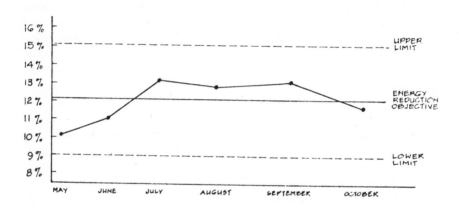

Control by inspection, then, is a warning control technique. It is used in the third step of supervision — checking on performance. Remember that control by inspection requires that boundaries and checkpoints be established during the planning activity.

Management, of course, is a more complex process than the foregoing discussion suggests. However, observance of those fundamental principles will go a long way toward ensuring the successful implementation of an effective energy conservation program. ·

ESTABLISHING PLANT PRIORITIES

Modification of energized systems, such as boilers and chillers, and nonenergized systems, such as controls and building envelope, will result in energy savings, but the responsibility for developing, implementing, and sustaining an effective energy management program rests with a third component, which can be called the human system. No matter how much time, effort, and ingenuity is expended in quick fix, refit, and systems convert modifications, an energy management program cannot be successful or effective without the understanding and support of all of the plant's employees.

The cost of energy is an important criterion in determining priorities within the plant. All plant employees will have to alter their personal energy-consuming habits in order to cope with escalating energy costs. The managers can capitalize on their increased energy awareness by drawing parallels between personal sacrifice at home and conservation efforts at work.

Experience has shown that the effectiveness of an energy management program frequently depends on the ability of the plant manager and chief business officer to work together on the development and implementation of an energy management plan. This collaboration helps to ensure that the program is given proper technical direction and financial support and that an energy dialogue is developed.

After this cooperative atmosphere has been established and energy cost and consumption data have been collected and analyzed, program priorities will begin to emerge. The highest priority should be given to options presenting the greatest return in energy savings in the shortest amount of time using the least amount of capital — the quick-fix projects. Records are kept of what quick-fix changes are made, the amount of energy saved, and the return on investment. After these modifications are implemented, a broader based, longer range plan should be formulated incorporating refit and, in some cases, systems convert options. Specific goals, the mechanisms and persons to be employed to achieve them, the amount of energy and dollars saved, and the payback period involved should be identified clearly in this broader plan.

Any effective energy management plan also must have a degree of flexibility built into it because plant priorities, energy supply, and energy management technologies change rapidly. Accordingly, an ability to respond to these changes, while moving forward in the effort to eliminate energy waste, mandates the development and administration of a carefully calculated plan that at the same time is broad enough to handle a variety of situations.

SETTING GOALS

The importance of setting clearly stated and reasonably attainable energy consumption reduction goals and working toward them cannot be overemphasized. Such goals can be set only after careful examination of cost and consumption data. An understanding of quick-fix potential and a knowledge of funds available for energy management investment also are necessary before setting goals.

Realistic goals can be achieved by developing a step-by-step plan of action that not only addresses the modification of energized and nonenergized systems but also takes into account the needs and idiosyncracies of the employee involved. One

Table 10-7 Energy Management Program

Location	Description	Estimated cost	Estimated annual cost	Payback period	Estimated life of installation	Estimated ROI	Date of modification	Personnel responsible
Canning Line A	Modify existing heating system and controls	$15,000	$9,000	1 year, 8 months	12 years	51.7%		
Cafeteria	Split Econostat zone and balance heating system..........	4,300	1,700	2 years 6 months	12 years	31.2%		
Canning Plant - Bldg. C	Modernize existing heating system and install automatic controls	9,000	1,700	5 years, 4 months	15 years	12.2%		
Business Office- Bldg. A	Service, repair or replace steam valves and traps	12,000	3,000	4 years	5 years	5 %		

helpful method for setting forth goals and plans of action to achieve these goals is the use of a clear and concise format for displaying energy management modification and anticipated results. A sample format is shown as Table 10-7. This format points out clearly the facility or area to be modified; the type, extent, and cost of the modification, and the energy savings and payback period.

Other important data displayed are the life of the modification or installation, the return on investment, the date of the modification, and the person responsible for the modification.

This step-by-step action plan, in simple yet complete fashion, will benefit everyone involved in the program and will serve as an energy management program guide that will be of use in checking progress and monitoring performance.

Although detailed plans may be developed for one year at a time, future years should not be ignored. Because energy management is an evolutionary process, effective program development and implementation require time and evaluation. An energy management program cannot be developed overnight by the sudden application of sophisticated computer systems or other capital-intensive substitutes for human effort.

Given the need for careful analysis and evaluation and the fact that a well-coordinated energy management program will yield significant short-term returns, an overall plan of energy management action probably should be developed spanning a relatively short period of time — approximately three years. A three-year plan is indicated in Figure 10-6.

Figure 10-6 Three-Year Energy Management Plan

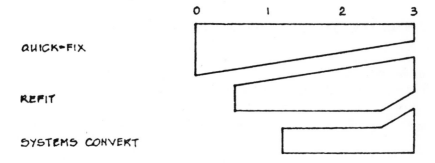

During the first year, it is important to concentrate on quick-fix measures as plans for refit and systems convert options are analyzed. The second year of a plan should include the implementation of specific refit measures along with further quick-fix measures, while the third year would focus on systems convert options for certain institutions.

Effective energy management is both an evolutionary and a simultaneous process. That is, even as progress is made from one phase to the next, implementation of conservation measures in all three phases will be taking place. For example, as refitting of HVAC systems occurs, there also may be quick-fixing of domestic hot water systems and removal of excess lamps. No one phase ever is finished completely. Each year, work in all three phases forms a part of the comprehensive program.

GAINING UNDERSTANDING, COMMITMENT, AND SUPPORT

Because an effective energy management program affects the entire facility and because it requires the allocation of capital resources, it cannot be successful without the understanding, commitment, and support of the top management team. The pace and tone of a successful program must be set by those who control the organization.

This type of understanding and support will depend on sound energy management planning and accurate identification of energy-saving modifications having the potential for rapid return on investment. Accordingly, the plant manager and the chief business officer should make certain that carefully constructed plans are developed and presented to the president.

EMPLOYEE COMMUNICATION AND MOTIVATION

At one time or another, employees of a plant are affected by an energy management program, so it is advisable to keep everyone fully informed of program goals and progress and to ask for advice where appropriate. Insights, ideas, and co-operation can be gained from employees if they are regarded as resources and are invited to participate. An energy committee often is a convenient and beneficial way to involve the entire staff. The committee can inform employees why certain measures are taken and advise them what to expect from such steps. By sharing and asking for plans and ideas, understanding and support for management energy programs can be generated.

Supervisors are concerned about any program that affects their departments, and thus it is important to gain their understanding, commitment, and support. Information on the energy management program should be available to them and briefings should be provided. Expert, low-cost advice often can be obtained from supervisors, particularly in the engineering department, with the concurrent benefit of winning support for the program.

Operating and maintenance personnel as well as technical and clerical staff also can provide assistance in an energy management program. Suggestions may be solicited and support urged so that this group of energy users is made to feel actively involved in the program.

ORGANIZING THE REVIEW PROCESS

Recommendations from all groups of plant personnel must be given careful consideration. It is vitally important not to overlook the special interests of any one group. Incorporation of worthwhile suggestions into the plan will make it that much more responsive to, and therefore supportable by, those it is intended to serve.

The goals of the energy conservation program must be kept in proper perspective with respect to the entire organization. An energy budget may range from 3 percent to 8 percent of the total budget, whereas personnel costs are a much higher percentage. If energy conservation activities impair the productivity of workers performing their assigned duties, they can lose their cost-effectiveness quickly. A reasonable balance must be maintained in order to achieve cost savings goals.

Joint Committee

It may be advisable to ask a group of selected administrators, supervisors, and employees to serve on a continuing energy management advisory committee. A joint committee can help gain the type of commitment and support necessary to implement energy-conserving measures in all areas of the plant. It would be easier for the head of a given department to "advise" the staff that the offices in the general office building will not be heated on weekends than it would be for the maintenance office to make the same statement. Many plants find it effective to have the chief administrator appoint such a committee. This not only invests the group with the type of credibility it needs but also provides opportunity to involve the president further.

Energy Management and the Facility

The way in which an energy management program is presented is important. Since in many cases schedules are to be altered, temperatures raised or lowered, lights removed, funds expended and other energy- and cost-saving measures implemented that directly affect the community, care must be taken to convey the need for and results of such actions. Such an explanation or presentation not only will assist those responsible for managing the program in gaining the understanding of the community, but also will generate support and enthusiasm for the program.

Official Announcement

To provide the level of support necessary to generate awareness, many institutions have announced energy management program plans through the office of the chief administrator. A representative public announcement of the initiation of an energy management program probably would include such items as the reasons for the program, planned consumption reduction goals, and the types of modifications to be implemented, as well as the facility's position on such issues as temperatures, vacations, and scheduling. Any such announcement also should stress that saving energy is synonymous with saving money — money needed for the employees and the programs. This point can be underscored by citing energy price increases over the last five years and the impact of the boosts on the institution's budget. Ultimately, all of the reasons behind an official public announcement of an energy management program relate to the need to focus attention on the energy issue and to inform the community on what is being done.

Continuing Communications

The need for continuing communication with the general plant after the program has been launched is critical. Managers and other employees must be kept aware of program progress, problem areas, measures to be implemented, and the amount of energy and money saved. Interest in and support for the program will fade quickly without this communication, and conservation progress can suffer as a result.

Newsletters and Brochures

Development and issuance of an energy newsletter and/or brochure explaining the energy management program will be valuable. Existing resources such as staff newsletters and weekly calendars also can be used to convey program progress and conservation tips as well as to give credit to those who have contributed significantly to the program. Brochures need not be limited to one time only; they may be issued periodically. Typically they highlight energy conservation activities, progress and plans for the near- and long-term future. They also become a permanent and valuable record of the energy management program.

Local Announcements

When possible, energy facts and figures for a specific facility should be made available to its personnel. The information fosters interest as well as involvement on the part of those responsible for the facilities or departments. Some plants have found that bulletin boards or large charts displaying progress in the reduction of energy consumption over time are useful. Strategically placed, the charts serve as continuing reminders of the program while providing visible incentives for continued conservation of fuel and money.

Major Policy Announcements

Energy conservation technology, as well as energy supply and demand, changes constantly. Foreign control of oil resources, development of alternate energy sources, and the needs of the plant itself mandate an energy management program that is capable of changing as consumption reduction goals are pursued or redefined. Occasionally, situations may require either a relatively radical change in the energy management program or new conservation guidelines in order to maintain progress. Such major policy decisions or changes should be conveyed directly to employees by the appropriate supervisor. For example, if the administration decides to prohibit the purchase of unit air-conditioners, a policy announcement to this effect must be issued. Such a decision requires the endorsement and support of a senior administrator because of its mandatory nature and its potential controversy.

Theme and Logo

Development of a theme and/or logo to characterize an energy management program will help gain program recognition and sustain enthusiasm for energy conservation efforts. A short phrase and/or an eyecatching logo will draw attention and give identity to the energy management program. Phrases and logos can be used for a variety of purposes: promotional brochures, stationery, posters, and bulletin boards. The more people who see the name and logo of the program, the more attention the program will draw. The more attention the program draws, the greater the likelihood that employees will become interested and involved.

Congratulations

Persons who have made significant contributions to the program should be congratulated. Giving praise when due is important in the effort to gain and maintain support.

Additional Ideas

There are a number of additional promotion and recognition devices worth noting. In most cases, such supplementary aids are inexpensive, yet their use will prove beneficial in an energy management program.

Point-of-Use Materials

Point-of-use materials are labels and signs that can be affixed in appropriate places to remind employees to take certain steps or to let them know certain steps have been taken. Stick-on light switch labels can be used to remind workers to turn off lights when not in use. Signs can be put in showers urging less use of hot water or to remind users that hot water temperatures have been reduced to conserve energy.

SUMMARY

Many ideas have proved successful in organizing and managing energy programs at various plants, but not all the measures mentioned here will be applicable to all facilities. There are, however, certain fundamental criteria for ensuring the success of an energy program:

- Teamwork is critical; there must be collaboration between the offices of the plant engineer and the chief business officer for a program to be successful, and staff must be organized for maximum effectiveness.
- Commitment from the president and support from the employees will be instrumental in ensuring progress.
- Goals and objectives that are stated clearly will be necessary for keeping everyone aware of the design and direction of the program and will allow evaluation of program progress and results.
- Communication to and from all employees will be critical to provide proper service to the facility, to understand its needs and problems, and, most importantly, to gain the support needed to make a program successful.
- Evolutionary progress is the key to energy management success. To be effective, the program should move forward carefully and methodically so that energy savings can be attained and maintained.

In addition to these fundamental criteria, it may be helpful to adopt some of the suggestions given to develop, sustain, or improve an energy management program. As the program progresses, flexibility also will be required: a willingness to alter priorities and pursue new opportunities, to consider special requests, to realize that all goals may not be achievable, or that new ones are in order. It will be necessary to change direction or to reassess efforts occasionally so that maximum energy savings can be achieved without compromising the business mission.

Chapter 11

Alternate, Nondepleting Energy Sources

SOLAR ENERGY

Each day the sunlit side of the earth receives the energy equivalent of 4,000 trillion kilowatt-hours. By contrast, the whole world uses roughly 80 trillion kilowatt-hours per year or .3 trillion kwh a day. Solar energy is available to all. It cannot be cut off by blockade or embargo. Nor does it involve the burning of any scarce fuels or the generation of any pollution. It cannot be exhausted, for it will last as long as the sun shines.

With this potential, it is no wonder that people are excited by solar energy. However, there are disadvantages. While the total amount of solar energy is great, it is spread over a vast area. To be useful, it must be collected and concentrated. This requires a great deal of space for mirrors or collectors. Solar energy also is subject to interruption. It is not available on cloudy days or at night. To ensure a dependable supply, the captured energy must be stored in some way or supplemented with a backup system. In addition, solar energy is not as flexible in its application as are more conventional sources of power. A fuel tank or an electric motor can be placed almost anywhere, but to use solar energy, space in which to spread solar collectors is needed in a sunny place. As long as fuel is cheap, there is no particular reason to suffer these disadvantages for the sake of using solar energy. Now that the cost of energy is rising sharply, solar energy is getting much more attention.

The availability of solar energy varies from place to place according to the prevailing weather conditions. In the Southwest, with its high percentage of clear days and nights, solar energy is at its best. In areas such as the Pacific Northwest or northern New York state, it is limited because of frequent clouds and overcast. An almanac or atlas, or Weather Bureau records, can provide information on how much sun can be expected in any area each season.

Solar radiation is measured in langleys, a unit named after Samuel P. Langley, who invented instruments for measuring that phenomenon. One langley is the radiation energy equivalent to one calorie falling on an area of one square centimeter. The measurements are taken on a horizontal surface at ground level.

The solar radiation flux varies from zero to 1.5 langleys per minute. One langley per minute is a typical value to expect on a clear day. Since the sun's radiation is measured on a horizontal surface, the experimenter can expect a bit more when tilting the collector toward the sun so that it is perpendicular to the sun's rays.

A solar collector is a device to absorb heat from the sun. Any object left lying in the sun will do this to some extent, but a solar collector is designed to collect and transfer as much heat as possible with minimum losses. The heart of a collector consists of coils or channels containing a fluid (usually air or water) so the absorbed heat may be transferred easily to where it will be used. By focusing the sun's rays with lenses or mirrors, much higher temperatures can be attained. Such a focusing type collector must follow the sun closely. It is most effective on clear days. On a hazy day, the sun's rays are scattered and the focusing collector, which "sees" only a small part of the sky, can pick up only a small percentage of them.

The alternative to the focusing collector is the flat plate collector. The fluid-filled coils or channels are spread flat and enclosed in a shallow box. The box is insulated and covered with glass to cut heat losses. It is best at supplying large amounts of heat at low temperatures. While the flat plate collector does not reach such high temperatures as the focusing collector, it does not need a sun-tracking mechanism. Since it absorbs radiation from the entire sky, it works well even on hazy days. In addition, it is more rugged and easier to construct than the focusing collector. For these reasons, the flat plate collector is the one used most often, except when very high temperatures are needed.

A focusing collector is aimed directly at the sun, while a flat plate collector usually is fixed in position. Since the height of the sun's path across the southern sky varies with the seasons, the tilt of the collector must be a compromise. The sun is highest in summer, lowest in winter, and midway between during spring and fall. The designer can take advantage of this by tilting the collector so it faces the sun squarely during the time of year when energy is needed most. For instance, a collector in a solar home heating system will be tilted to face the low winter sun so that the device will work best during the coldest part of the year. The coils of a collector nearly always are blackened to make them more efficient at absorbing light and converting it to heat. Unfortunately, black also is good at radiating heat, and most of this re-radiated heat is lost. To stop this, some collectors use selective surfaces. These are coatings that are good at absorbing light, but poor at radiating heat. The use of selective surfaces makes a collector much more efficient. Unfortunately, the making of selective surfaces is beyond the amateur at present.

The amount of energy gathered by a collector depends to some extent on the flow rate of the heat transfer fluid through it. If the rate of flow is slow, the fluid has plenty of time to reach a fairly high temperature. On the other hand, the hotter the collector is allowed to get, the more heat it will lose to its surroundings. When the rate of flow is fast, the fluid does not warm up much. However, a larger amount of fluid is warmed, and since the losses are less in this mode, the total amount of energy delivered by the collector is greater.

Most applications of solar energy require some means of storing the heat gathered. This can be as simple as a large tank of water. Other methods include bins of stones and stacks of water-filled jugs around which warm air is allowed to circulate. These methods all make use of the capacity of a large bulk of material to absorb and retain heat. Another method depends on fusion heat. Materials are capable of absorbing large amounts of heat as they melt and then releasing it as they resolidify. Sodium sulfate, or Glauber's salt, has been used for this purpose. It melts at 32° C (77° F). By using other salts and combining them in various proportions, it is possible to get mixtures with lower melting points. Such mixtures are called eutectic. Systems using molten salts have the advantage of compactness, though they are more expensive than the other methods mentioned.

In the long term, solar energy almost certainly will be one of the major sources of energy. In the short term, solar energy can be used to stretch supplies of fossil fuel. Right now large amounts of fossil fuels are burned to provide low temperature heat, as in home heating. This could be done just as well by solar energy, which would free large amounts of fuel for tasks requiring concentrated high temperature heat.

Sun and Sky Radiation for Heating and Cooling

There are three technological processes by which solar energy can be utilized: (1) heliochemical, (2) helioelectrical, and (3) heliothermal. The first process, through photosynthesis, maintains life on this planet; the second, using photovoltaic converters, provides power for all of the communication satellites and ultimately will be valuable for terrestrial applications; the third can provide much of the thermal energy needed for space heating and cooling and hot water heating.

Solar Heat Collection

Solar radiation data may be used to estimate how much energy is likely to be available to be collected at any specific location, date, and time of day either by a concentrating device, which can use only the direct rays of the sun, or by a flat plate collector, which can use both the direct and diffuse radiation. Since the temperatures needed for space and cooling are quite moderate, not exceeding 250° F even for absorption refrigeration, they can be attained by carefully designed flat plate collectors. A flat plate collector generally consists of five components:

(1) glazing, which may be one or more sheets of glass or a diathermanous (radiation-transmitting) plastic film or sheet
(2) tubes or fins for conducting or directing the heat-transfer fluid from the inlet duct or header (two-in) to the outlet (two-out)
(3) plate, generally metallic, which may be flat, corrugated, or grooved, to which the tubes or fins are attached in a manner that produces a good

thermal bond
(4) insulation, which minimizes downward heat loss from the plate
(5) container or casing, which surrounds the foregoing components and keeps them free from dust and moisture.

During the last century, flat plate collectors have been constructed from many materials and in a wide variety of designs. They have been used to heat water, water plus an antifreeze additive such as ethylene glycol, water plus ammonia or other refrigerants, fluorinated hydrocarbons, air, and other gases. The major objective has been to collect as much solar radiation as possible, at the highest attainable temperature, for the lowest possible investment in labor and materials. When these goals are attained, the collector should have an effective life of many years despite the adverse effects of the sun's ultraviolet radiation, corrosion or clogging due to acidity, alkalinity or hardness of the heat-transfer fluid, freezing and air-binding in the case of water or depositing of dust and moisture in the case of air, and breakage of the glazing due to thermal expansion, hail, or other causes.

Glazing Materials

The purpose of glazing is to admit as much solar radiation as possible and to reduce the upward loss of heat to the lowest attainable value. Glass has been the principal material used to glaze solar collectors because it has the highly desirable property of transmitting as much as 90 percent of the incoming shortwave solar radiation, while virtually none of the longwave radiation emitted by the flat plate can escape outward by transmission. Glass of low iron content has a relatively high transmittance (approximately 0.85 to 0.90 at normal incidence) for the solar spectrum from 0.30 to 3.0M, but its transmittance is essentially zero for the longwave thermal radiation that is emitted by sun-heated surfaces (3.0 to 50M).

Plastic films and sheets also possess high shortwave transmittance but, because most of the usable varieties have transmission bands in the middle of the thermal radiation spectrum, they may have longwave transmittances as high as 0.40.

Plastics generally are limited in the temperatures they can sustain without undergoing dimensional changes, and only a few of the varieties now available can withstand the sun's ultraviolet radiation for long periods of time. They possess the advantage of being able to withstand hail and other stones and, in the form of thin films, are completely flexible.

The glass generally used in solar collectors may be either single strength (0.085 to 0.100 in. thick) or double strength (0.115 to 0.133 in.) and the commercially available grades of window or greenhouse glass will have normal incidence transmittances of about 0.87 and 0.85 respectively. For direct radiation, the transmittance varies to a marked extent with the angle of incidence.

For clear glass such as that used for solar collectors, the 4 percent reflectance from each glass-air interface is the most important factor in reducing transmission, although about 3 percent gain in transmittance can be obtained through the use of "water white" glass. Antireflection coatings of the kind used for camera and telescope lenses also can make significant improvement in transmission, but the cost of the processes now available is prohibitively high.

The effect of dirt and dust on collector glazing is surprisingly small, and the cleansing effect of occasional rain seems adequate to maintain the transmittance within 2 to 4 percent of its maximum value.

Glass is virtually opaque to the longwave radiation emitted by the collector plate, but the absorption of that radiation causes the glass temperature to rise and thus to lose heat to the surrounding atmosphere. This type of heat loss can be reduced by using an infrared-reflecting coating on the underside of the glass, but such coatings are costly and also reduce the effective transmittance of the glass for solar radiation by as much as 10 percent.

In addition to serving as a heat trap by admitting shortwave solar radiation and retaining longwave thermal radiation, the glazing also reduces heat loss by convection. The insulating effects of the glazing are enhanced by the use of several sheets of glass, or glass plus plastic. The upward heat loss may be expressed by:

Qup = Acp $\times$ Ucp $\times$ (tp-to) BTUh (17) where Qup designates the heat loss upward from the collector plate of area Acp sq. ft., Ucp is the upward heat loss coefficient in BTUh/ft^{2}F, and tp and to are the temperatures of the collector plate and the ambient air, respectively.

Collector Plates

The primary function of the collector plate is to absorb as much as possible of the radiation reaching it through the glazing, to lose as little heat possible upward to the atmosphere and downward through the back of the container, and to transfer the retained heat to the transport fluid. The absorptivity of the collector surface for shortwave solar radiation depends upon the nature and color of the coating and upon the incident angle.

Prior to 1955, flat black oil-based paint was the universal choice for coating solar collectors. The publication of papers by Tabor of Israel and Gier and Dunkle of the United States led to awareness of the fact that there was virtually no overlapping of the shortwave solar spectrum (0.3 to 3.0M) and the longwave thermal spectrum (3.0 to 30M).[1][2] By suitable electrolytic or chemical treatments, it is possible to produce surfaces that have high values of solar radiation absorbtance and low values of longwave emittance. Essentially, such "selective surfaces" consist of a very thin upper layer that is highly absorbent to shortwave solar radiation, but relatively transparent to longwave thermal radiation. The substrate must have a high reflectance and a low emittance for longwave radiation.

Among the selective surfaces that have demonstrated their ability to retain their desirable properties after long exposure to intense sunshine are those produced by Tabor's electrolytic processes and used in the mass production of solar water heaters in Israel, with galvanized sheet steel as the plate material. Solar absorptivities in the range of 0.92 and longwave emittance as low as 0.10 characterize the Tabor surfaces.

F. Daniels, who authored *Direct Use of the Sun's Energy,* published by Yale University Press in 1964, gave a comprehensive analysis of the chemical aspects of producing selective surfaces and Christie gave details of current Australian practices in this important field.[3]

Selective surfaces are of particular importance when the collector surface temperature is much higher than the ambient air temperature. The references cited above are only a few of the large number of publications on this subject, many as a result of the use of selective surfaces in the space program.

If the objective of the collector is to heat liquids, tubes must be integral with or attached to the plate with a good thermal bond. Beginning with the classic work of Hottel and Woertz at M.I.T. and continuing with the studies by Whillier and Erway, many investigations, both theoretical and experimental, have been made into the details of collector design and performance.[4] [5] The method of attaching the tubes to the collector plate has challenged the ingenuity of designers in all parts of the world. The major problem is to obtain a good thermal bond without incurring excessive costs for labor or materials.

Materials most frequently used for collector plates, in decreasing order of cost and thermal conductivity, are copper, aluminum, and steel. If the entire collector area is swept by the heat transfer fluid, the conductivity of the material becomes unimportant. Whillier's study of the effect of bond conductance concluded that steel pipes were just as good as copper if the bond conductance between tube and plate was good. Bond conductance can range from a high of 1,000 BTUh/ft/F for a securely soldered tube to a low of 3.2 for a poorly clamped or badly soldered tube. Bonded plates with integral tubes are among the best alternatives as far as performance is concerned, but they require mass production facilities.

Forced Water Circulation

There are many applications where the capacity of a thermosyphon system with one or two collector panels is inadequate to meet an existing demand for hot water. In other situations, architectural or other considerations require that the storage tank be below or at a considerable distance from the collectors. In these cases, forced circulation systems are employed, with flow rates on the order of 1 gallon/hr per sq. ft. of absorbant surface. Centrifugal hot water accelerator pumps have proved satisfactory at these relatively low flow rates.

When a number of collector units are to be operated simultaneously, series-parallel arrangements have been found superior to operation with all of the collectors in parallel. Australian authorities report that up to 24 tubes may be operated in parallel; for larger numbers, downcomers or multiple parallel system may be used. Vents to avoid air binding are essential.

Forced Air Circulation

Solar air heaters have not yet received the detailed study that has been given to water heaters, but there are many applications where air is a more suitable heat transport fluid than water. Air cannot freeze, and while air leakage is annoying, it is not as serious in its consequences as are water leaks. Natural circulation of sun-heated air is being employed in the French houses at Font Remeu, near the great solar furnace at Odeillo. Since the specific heat of air is only 0.24 BTU/lb/F as compared with water's 1.0 BTU/lb/F, and its density is only about 0.075 lb/ft^{22} as compared with water's 62.4 lb/ft^{22}, it is obvious that the ducts needed to convey air to and from a solar collector must be far larger than the pipes used to carry water. However, most air-conditioning systems rely upon air to do the actual heating and cooling, and so air has some notable advantages as a heat transport medium.

Collection Efficiencies

The efficiency of a solar collector may be defined as the ratio of the amount of heat usefully collected to the total solar irradiation during the period under consideration. Instantaneous efficiencies during the middle of the day, when the incident angle is favorable, generally are higher than day-long efficiencies that must take into account the high and unfavorable incident angles that prevail during early morning and late afternoon. A highly simplified approach to this complex subject may be presented in the following terms:

Collector heat gain, = BTUh/ft^2	Heat transport fluid flow rate Lb/hr/sq ft	Specific × heat of fluid	temperature × rise of fluid, deg F

A heat balance, expressed in terms of unit area of collector surface, may be stated as:

Solar irradiation rate	Collector = heat gain rate	Solar radiation dissipated from + glazing to sky miscellaneous heat losses from casing and piping	Heat loss + upward from collector

The loss due to reflection from the cover glass or glasses is approximately 4 percent of the energy passing through each air-glass interface for incident angles up to 35 degrees, so 8 percent is lost for a single-glazed collector, 15 percent for double glazing, and 22 percent for triple glazing.

Disposition of the energy absorbed by the glazing is more complex, since a rise in the glass temperature means that the heat flow from plate to glass is reduced somewhat. When absorption is taken into account, the losses for incident angles up to nearly 40 degrees are: for single glazing, 10 percent; for double glazing, 18 percent; for triple glazing, 25 percent. The use of thin plastic films would reduce these losses significantly, but such films have the disadvantage of transmitting 30 to 40 percent of the longwave thermal radiation emitted from the collector plate. The combination of an outer glass plate and one or more inner plastic films has interesting possibilities.

The major component of the losses from a well-insulated collector is the upward heat flow from the plate to the atmosphere. The upward heat flow is a function of the emittance of the plates for longwave radiation, the temperature difference between the plate and the air above the glazing, and the wind velocity.

The amount of radiant energy absorbed by the collector plate is the product of the incident irradiation, Its, the transmittance of the glazing, t_g. and the plate's absorbtance for solar radiation, a_s. The absorbtivity usually is well above 90 percent for incident angles up to about 40 degrees, but above that angle both the absorbtivity and the transmittance drop off rapidly.

For very low fluid temperatures, which may be below the temperature of the ambient air, the efficiency is highest without any cover glass.

As the absorber temperature rises, the efficiency drops off rapidly with a single cover glass and a nonselective absorber. The addition of a good selective surface, $e_s = 0.10$, makes a marked improvement in efficiency as temperatures rise toward the 200° F level, where the nonselective collector finds that the reduced upward heat flow coefficient of double glazing is just offset by its reduced transmittance. At a collector temperature of 250° F, desirable for conventional absorption refrigeration, 40 percent efficiency can be attained with a triple-glazed high emittance collector or with a single-glazed collector $e_s = 0.10$.

Concentrating Collectors

Temperatures far above those attainable by flat plate collectors can be reached if a large amount of solar radiation is concentrated upon a relatively small collection area. The use of simple flat reflecting wings can increase the amount of direct radiation reaching a collecting surface to a marked extent, but because of the apparent movement of the sun across the sky, concentrating collectors must be able to "follow the sun" during its daily motion.

Reflective troughs with parabolic cross sections running east and west require the minimum amount of adjustment since they need to be corrected only for changes in the sun's declination. There inevitably is some degree of morning and afternoon shading of the ends of the collection tube, which generally is mounted along the focal line of the parabolic trough.

Paraboloidal concentrators, resembling searchlight reflectors, can attain extremely high temperatures, but they require very accurate tracking systems and can use only the direct rays of the sun, since diffuse radiation cannot be concentrated. There are two primary methods of tracking the sun. The altazimuth method requires that the tracking surface change both its altitude and its azimuth to follow the sun in its motion across the sky. In the equatorial mounting method, the axis of the concentrator is pointed to the north and arranged to change its tilt angle to compensate for the varying declination of the sun. The daily motion is simply rotation at the rate of 15 degrees per hour to compensate for the sun's apparent motion at the same angular rate. Good possibilities exist in the horizontal parabolic trough, oriented east and west, in which a multiplicity of small troughs are mounted in a fixed frame, inclined at the angle of the local latitude. Very little adjustment is needed if the absorbing pipe is relatively large. E.A. Farber also reports the use of an equatorially-mounted parabolic trough that intercepts a 6 ft.-by-8 ft. beam of solar radiation and heats fluid to about 600° F.[6]

The principal use of concentrating collectors has been in the production of steam or high temperature fluids for use in refrigeration or power generation. The higher cost and added mechanical complexity of collectors that must follow the sun, and their inability to function at all on cloudy or overcast days, are disadvantages that must be borne in mind.

Cooling by Solar Energy

The use of heat to produce cold has been known for well over a century. In Paris in 1878 some steam produced by a solar boiler was used to operate a primitive absorption refrigerator and produce a small quantity of ice. Since that time, other experimenters have explored three processes by which the sun's radiant energy can produce cooling effect. All three systems are used widely with conventional energy sources.

The first is the steam jet system, investigated in 1936 by W.P. Green at the University of Florida.[7] A concentrating collector was used to produce steam at a pressure high enough to make a steam jet ejector function that in turn caused evaporation and chilling of water in a tank connected to the ejector. The Coefficient of Performance proved to be low and the requirement for a sun-following concentrator made the system impractical from a commercial point of view.

The second system employs the familiar compression refrigeration cycle, driven by a Rankine cycle engine or turbine, operated by steam or some other vapor that can be generated in a solar collector.

The availability of fluorinated hydrocarbons as working fluids has reawakened interest in solar-powered compression refrigeration. Teagen and Sargent have proposed an ingenious system in which refrigerant-type working fluids, particularly R-114, will be used for the power cycle and R-22 in the refrigeration cycle.[8] One version of their system uses a four-cylinder reciprocating engine, with two cylinders providing the power while the two others compress the R-22. They envision a system that also can produce electric power in winter when cooling is not needed. With a solar collector temperature of 120° C (248° F) and an air-cooled condenser, they aspire to a Coefficient of Performance as high as 0.4, that, while low compared to motor-driven compression systems, is good in comparison with absorption systems.

Löf discusses the operation of commercially available absorption refrigeration systems using water vapor as the refrigerant and lithium bromide solution as the absorbent.[9] Tests conducted at the University of Wisconsin showed that hot water from a solar collector at 175° F could produce two tons of refrigerating effect from a commercial unit rated at three tons when supplied with heat at 250° F. Despite the lowered rating caused by the reduced temperature, the Coefficient of Performance remained at 0.6.

The University of Florida has concentrated the attention of its Solar Energy Laboratory upon the ammonia-water absorption cycle. Farber reports the successful operation of a continuous refrigeration system in which the solar absorber also is the generator that drives the ammonia out of the water solution. The Florida systems provide cooling with lower collector temperatures than any of the other methods thus far reported, and appear well adapted to use with the type of flat plate collectors that can be built today.

Intensive research will be needed to produce solar air-conditioning systems that will be low enough in cost and high enough in performance to compete with conventional systems, but solar cooling has the great advantage of having the largest supply of energy available when the demand is highest.

F.S. Dubin in his testimony before the Energy Subcommittee of the House Committee on Science and Technology discussed solar heating and cooling of large office buildings.[10] He noted that absorption equipment now was available in the 100-ton range that could operate satisfactorily at water temperatures of 210° F to 220° F, rather than 270° water or 15 psig steam. He anticipated the development of absorption units operating successfully at 180° to 190° F. Farber reports that this has been accomplished already with units of a size suitable for single-family residence, where the roof area generally is more than adequate for the collectors.

WIND POWER

Production of energy by wind power makes solar-reflecting devices insignificant in comparison. The inventions that impound solar energy through various lensing devices are fascinating to the imagination but they work only a few hours daily when the sun is at a favorable angle. Even if half of Arizona were turned into a direct-sunlight converting mechanism, the production of energy would be negligible in comparison with wind power sources.

Wind power is in a class by itself as the greatest terrestrial medium for harvesting, harnessing, and conserving solar energy. The water and air waves circulating around the earth are unsurpassed energy accumulators whose captured energy may be used to generate electrical, pneumatic, and hydraulic power systems.

Windmills produce power from the sun-generated differentials of heat (which are the source of all wind) with far greater efficiency than do attempts to focus and store direct solar radiation. But the most comprehensive consideration regarding wind power is not technological. Rather, it is an appreciation that wind power is by far the most efficient way to recapture solar power.

Three-quarters of the earth is covered with water and the remaining quarter is land area consisting largely of desert, ice, and mountains. Only about 10 percent of the planet's area has terrain suitable to cultivation, in which vegetation can impound the sun's radiation by photosynthesis.

Among the solar energy impounders in vegetation, none can match corn's performance. Corn converts and stores as recoverable energy 25 percent of the received ultraviolet radiation, whereas wheat and rice average only 18 to 20 percent. From these stores of solar energy, humans can produce commercial alcohol or they can leave the energy to the production of fossil fuels in the earth's crust, which requires millennia.

But one-half of the vegetation-producing area of the planet's surface always is in the shadow, or night, side, which reduces to 5 percent the working area of the surface on which vegetation impounds the sun's energy. Theoretically, 5 percent of the area can impound energy at any one time, but an average of only one percent of the sun's energy actually is being converted because of local weather conditions and infrared and other energy-radiation interferences.

The area of the surface of a sphere is exactly four times the area of the sphere's great-circle disk, as produced by a plane cutting through the center of the sphere. The surface of a hemisphere is, then, twice the area of the sphere's great-circle plane. The "full" moon actually is a surface twice the area of the seemingly flat, circular disk in the sky.

All of the earth's energy comes from the stars, but primarily from a single star, the sun, as radiation or as interastrogravitational pull. Twenty-four hours a day the sun drenches the outside of the hemisphere of the cloud-islanded atmosphere's 100-million-square-mile surface area, which is twice that of the disk of the earth's profile.

This provides one billion cubic miles of expanding atmosphere on the sunny side and one billion cubic miles of contracting atmosphere on the shadow side. The atmospheric mass is accelerated kinetically in the hemisphere, which is constantly saturated by the sun, while simultaneously the atmospheric kinetics in the night atmosphere are decelerated.

All around the earth, yesterday's sun impoundment perturbates the atmosphere by thermal columns rising from the oceans and lands. Rotation of the earth brings about a myriad of high-low atmospheric differentials and world-around semivacuumized drafts, which produce the terrestrial turbulence called weather.

The combined two billion cubic miles of continual atmospheric kinetics converts the solar energy into wind power. Wind power is sun power at its greatest, by better than 99 to 1.

All biological life on earth is regenerated by star energy, and overwhelmingly by the sun's radiation. The sun radiates omnidirectionally 92 million miles away from the earth, with only two-billionths of its total radiation impinging upon this planet. The radiation arrives at a rate of two calories of energy per each square centimeter of earth's sunside hemispherical surface per each minute of time. About half of that is reflected back omnidirectionally to the universe. The other half, i.e., 221.216 BTU's per hour per square foot, is impounded by the planet's biosphere, thus making the energy available for human use.

No matter how dubious such logical realizations of these potentials seem, the fact remains that earth's net receipt and impoundment of cosmic energy amounts to 120 trillion horsepower, which can be stated also as 90 trillion kilowatts. With 8,760 hours a year, the total energy amounts to 786 quadrillion kilowatt-hours a year. This is 786×10^{15} kilowatt-hours, almost two hundred thousandfold the world's present 5×10^{12} kilowatt-hour production of electric energy.

If all humanity enjoyed 1978's "highest" living standard — that of the United States — each human on earth would consume 200,000 (2×10^5) kilocalories a day.

If it is assumed that there will be 5 billion (5×10^9) humans on the earth by the year 2000, with each person consuming 7.9365×10^5 BTUs daily, $1 \times 3.9682 \times 10^{15}$ BTUs per day will be required for the world's population. The actual daily income of solar energy is 7.9×10^{18} BTUs and the earth's usable solar energy per person per day is 2×10^3.

With the advent of rural electrification more than a third of a century ago, windmills were going out just as modern aerodynamic research was coming in. Recent years have seen the development of windmill-generated electricity. One of the strategies has been to convert pure sun-distilled water electrolytically into hydrogen directly for power purposes. The hydrogen and oxygen also could be reassociated to produce electric current at an overall 85 percent efficiency. A complementary strategy is the combination of improved, variable pitch windmill propeller blades with the aerodynamics of jet technology in which the windflow patterns embracing whole buildings are captured and funneled into low-pressure focusing, Pitot-tube cowlings. Also under development is a new octahedral windmill mast that is transportable, powerful, economical, and swiftly erectable, as well as a new low-cost method of mechanical linkage from the mill to the generator.

Experiments have found that the Greek-island type windmills with self-furling sails also are very efficient.

Experiments also show that flywheels, as energy accumulators, can be employed efficiently in connection with variable winds to drive generators.

The force of earth's winds produces 150 million square miles of ocean waves, forms clouds, causes violent storms, and fabricates a 200-billion-cubic-mile spherical mantle 100 miles thick that functions as a sun-energy storage battery adequate to accommodate and eternally regenerate energy for all of the world's needs and pleasures. The entire process is carried out with a safety factor coefficient of 10,000 to 1.

The U.S. Navy has stated that one minute of one hurricane releases more energy than that of the combined atomic bomb arsenals of the United States and the Soviet Union. From the viewpoint of design science, it is simply a matter of coping with the calm, zephyr, gale, or hurricane variabilities of wind power.

Wind Generator Basics

Virtually all electricity is produced by rotating generators that develop power by rotating magnets in front of each other. Power companies use huge generators turned by steam turbines or, in the case of hydroelectric power, by water turbines. The steam to turn the turbines is produced by boiling water over a fuel. Atomic energy power plants function in the same way except that the heat to produce the steam comes from radioactive fission instead of fossil fuels. In an automobile, the generator is turned through a V-belt by the gasoline engine. Similarly, in a small portable power plant, the generator is turned by a gasoline, diesel or low pressure gas engine.

Thus all forms of usable electricity come from some type of rotating generator driven by an external power source. The wind generator is no exception. It consists of a rotating generator turned by a propeller that in turn is pushed around by the force of the wind upon it. The propeller can be thought of as a wind engine using wind as its only fuel. The amount of electricity that can be generated by a wind generator is dependent on four things: (1) the amount of wind blowing on it, (2) the diameter of the propeller, (3) the size of the generator, (4) the efficiency of the whole system. The following are examples of how this works.

First, consider an 8-foot diameter propeller with well-designed blades having an efficiency of 70 percent and a generator capable of delivering 1,000 watts. In a 5 mph breeze, this might produce 10 watts of power; at 10 mph, about 75 watts; at 15 mph, 260 watts, and at 20 mph, 610 watts. In other words, the more wind, the more power. But it is not a simple relationship. The actual power available from the wind is proportional to the cube of the windspeed, so if the windspeed is doubled, it will produce eight times as much power.

Next, consider a propeller with a 16-foot diameter and a similiar efficiency to the first one. At 5 mph wind, it might produce 40 watts output; at 10 mph, 300 watts; at 15 mph, 1,040 watts; and at 20 mph, 2,440 watts if the generator were capable of delivering this much power. The power output of the 16-foot diameter windmill thus is about four times that of the 8-foot diameter windmill.

This shows that the power is proportional to the square of the diameter, or that doubling the size of the propeller will increase the output by a factor of four. These are the two basic relationships fundamental in the design of any wind-driven power plant. The table below illustrates these relationships.

Table 11-1 Windmill Power Output in Watts (assuming 70% efficiency)

Propeller diameter in feet	Wind velocity in mph					
	5	**10**	**15**	**20**	**25**	**30**
2	0.6	5	16	38	73	130
4	2	19	64	150	300	520
6	5	42	140	340	660	1,150
8	10	75	260	610	1,180	2,020
10	15	120	400	950	1,840	3,180
12	21	170	540	1,360	2,660	4,600
14	29	230	735	1,850	3,620	6,250
16	40	300	1,040	2,440	4,740	8,150
18	51	375	1,320	3,060	6,000	10,350
20	60	475	1,600	3,600	7,360	12,760
22	73	580	1,940	4,350	8,900	15,420
24	86	685	2,300	5,180	10,650	18,380

But what about efficiency and generator size? The efficiency (defined as the ratio of the power actually developed to the theoretical maximum power obtainable at a certain wind speed) depends largely on what type of propeller is used. All modern electric wind-generating plants use two or three long, slender, aerodynamically shaped blades resembling an aircraft propeller. These efficient propellers operate at a high tip speed ratio (the ratio of propeller tip speed to wind velocity). The Quirk's propeller, for example, runs at a tip speed ratio of about 6 while for some of the Swiss Elektro units the ratio runs as high as 8. This compares to ratios of 1 to 3 for the slower running multi-blade American water-pumping windmills. But while the latter type is less efficient, its higher starting torque and steadier speed at low wind velocities makes it more suited for pumping applications.

Ideally, the propeller of a wind machine used for generating electricity should have a cross-section resembling that of an aircraft wing, with a thick rounded leading edge tapering down to a sharp trailing edge. It should be noted, however, that the most efficient airfoils for aircraft propellers, helicopter blades, or fan blades (all designed to move air) are not the most efficient airfoils for windmills (which are intended to be moved by air). An old airplane propeller, in other words, has neither the proper contour nor angle of attack to extract energy satisfactorily from the wind. Those with the ability and time who wish to construct a propeller of their own will find several good designs in the United Nations publication, "Proceedings, Vol. 7, UN Conference on New Sources of Energy," 1971.

The generator itself forms the vital link between windpower and electrical power. Unfortunately, most generators that would seem to be suitable suffer from the requirement that they need to be operated at high speeds; they are built to be driven by gasoline engines at speeds from 1,800 rpm to 5,000 rpm. But windmill speed, especially in the larger sizes, seldom exceeds 300 rpm. This means resorting to special low-speed generators (which are expensive and cumbersome), or to some method of stepping up the speed of the generator using belts, sprockets, or gears. The large commercial units available generally present a compromise. They use a relatively low-speed generator (1,000 rpm) and gear the generator to the propeller through a small transmission at about a 5 to 1 step-up ratio.

The next question is, how to decide what size generator to use with what size propeller. Here again some compromises are in order. First, it must be decided what windspeed will be required for the generator to put out its full electrical output. If full output at low windspeeds is desired, a large propeller will be needed, whereas if full output only at high wind velocities is satisfactory, a small propeller will suffice. In general, light winds are more common than strong winds. Statistical studies of wind data (from the U.S. testing station in Dayton, Ohio) show that each month a well-defined group of wind velocities predominates. These are called the prevalent winds. There also is a well-defined group called energy winds that contains the bulk of the energy each month. The first group, consisting of 5 to 15 mph winds, blows 5 out of 7 days on the average, while the energy winds of 10 to 25 mph blow only 2 out of 7 days. It might seem logical to design for maximum output at 15 mph in order to take advantage of the prevailing winds, but this would require a very large propeller for the power produced, and all the power from winds higher than 15 mph would be lost. As an example, consider a 2,000-watt generator that yields its full output at 15 mph. From Table 11-1, it can be deduced (assuming a 70 percent efficient system) that to get 2,000 watts at 15 mph, a propeller larger than 22 feet in diameter will be needed. This will be large and expensive to build and difficult to control in high winds. Besides, much power is being thrown away at higher windspeeds. If a 22-foot-diameter rotor is to be built, then a bigger generator might as well be installed to get some of that power at higher windspeeds.

Most working generators are designed to put out full power in windspeeds of about 25 mph, but in so doing they sacrifice some performance at low windspeeds. Usually they deliver almost no output at winds below 6 or 8 mph, but this is not a serious drawback because there is so little energy available from these light winds.

This is what is meant by a "2,000-watt" wind generator. It is hard to compare the rated output of a wind generator to that of a conventional generating plant with the same rating. In the case of the wind generator, the power rating merely states the maximum output of the generator at a certain windspeed. This windspeed must be known to calculate how much power actually can be derived from a certain windplant under varying wind conditions.

The final question in choosing a suitable wind electric system is how much total electric energy a certain size system will produce over a period of time in any particular location. This is the main concern of those attempting to determine the feasibility of a wind generator in their area, and it also is the most difficult question to answer. Suppose 200 kilowatt-hours of electricity per month will be needed to run everything wanted with a new wind-powered system. If all the power is to come from the wind, a system will be necessary that will provide at least this much per month and a little more to allow for the slight inefficiency of the storage batteries.

To determine precisely how much a system will deliver in a given location, the complete output characteristics of the wind generator at different windspeeds must be known, as well as complete windspeed data for the proposed installation site (enough data to plot a continuous graph of the windspeed for a year or two). Such a graph would permit computing the total energy available from the wind in a particular location. Roughly, the power available would correspond to the area under the curve, but even this is not mathematically correct because of the cubic dependency of power on windspeed. To do it correctly would require sophisticated statistical analysis that actually is academic because few persons have adequate wind data for their locations. For more information on this, Putnam's book, "Power from the Wind" will be helpful.

In proceding to calculate the average yearly winds in a particular location must be ascertained. The Weather Bureau records windspeeds hourly at several hundred stations across the county and will provide this information, including average windspeeds for each month and year at a station near the plant. This is a start, but should not be considered definitive. Winds at the actual site may vary considerably from those at the local weather station, so you probably will want to carry out tests of your own, especially if you are in a doubtful area, i.e. official average winds much under 10 mph. Assuming that the average winds at your location are 12 mph, the following table will give some idea of what to expect from different size wind plants at various average windspeeds. Since many factors have entered into this, several assumptions have had to be made. First, these figures are based on typical present production wind generator designs with tip speed ratios on the order of 5 and efficiencies of about 70 percent. It also is assumed that there is negligible output below windspeeds of 6 mph and that maximum output is reached at 25 mph. Table 11-2 represents a composite of actual measurements, plus figures issued by several wind generator manufacturers, plus a fair amount of interpolation.

Table 11-2 Average Monthly Output in Kilowatt-Hours

Nominal output rating of generator in watts	Average Monthly Windspeed in mph					
	6	**8**	**10**	**12**	**14**	**16**
50	1.5	3	5	7	9	10
100	3	5	8	11	13	15
250	6	12	18	24	29	32
500	12	24	35	46	55	62
1,000	22	45	65	86	104	120
2,000	40	80	120	160	200	235
4,000	75	150	230	310	390	460
6,000	115	230	350	470	590	710
8,000	150	300	450	600	750	900
10,000	185	370	550	730	910	1,090
12,000	215	430	650	870	1,090	1,310

Table 11-2 should be considered as only a rough estimate of what can be expected from wind-generated power in different wind areas. Many manufacturers of wind generators refuse to commit themselves to anything as specific as the figures in this table. They contend that conditions vary so much, as do the effects of turbulence, temperature, etc., it would be difficult to make any specific predictions of long-term energy output. Nevertheless, this is the one basic statistic that everyone wants to know when considering installing a wind electric system.

Now, to use the table to solve the original problem: how large a system will be needed to get that 200 kwh per month in an area with a 12 mph average windspeed. The 12 mph column in Table 11-2 shows that a 2,000-watt system would produce only 160 kwh while a 4,000-watt generator would produce 310 kwh. Interpolating between these two values, it can be estimated that a 3,000-watt unit might produce 230 kwh per month, which is just about right, allowing for the inefficiencies of batteries, inverters, etc. Of course, when it comes to buying or building such a system, a company may be forced by financial or other considerations to install a larger or smaller system, but at least it will have some idea what to expect from it when the installation is completed.

Wind generators in high winds provide other types of situations. All modern production wind plants are designed to function completely automatically in winds up to 80 mph or more, so there is no such thing as a site with too much wind. To survive all kinds of conditions, wind generators employ some method of holding down their speed in heavy winds. The most common method of spilling excess wind, whenever the power from the wind exceeds the power rating of the generator, is a system of weights mounted on the propeller that act centrifugally to change the pitch of the blades, thus reducing the wind force on the propeller. This system, which serves as a built-in governor, holds the propeller at a constant speed and prevents overspeeding when there is little or no load on the generator. Such conditions occur whenever the batteries are fully charged and no power is needed. This is one area where the modern wind plant has come a long way in solving a problem that plagued the wind chargers of forty years ago. Burned-out bulbs and even burned-out generators were not uncommon with the old units as windmills raced out of control in heavy winds.

Even with the modern version, manufacturers generally recommend that if wind speeds greater than 80 mph are anticipated, as in a hurricane, the propeller should be stopped manually and/or rotated sideways to the wind. Most models have a brake control at the bottom of the tower for this purpose. Such "furling" of the windmill during a storm greatly reduces the strain of high wind loads on the propeller and on the entire tower structure.

SOLID-WASTE-BASED ENERGY

Conversion of solid wastes into energy is a process that merits serious consideration as other fuels become expensive and difficult to obtain. Each gallon or unit of energy costs more to find, mine, and refine. Nearby sources are becoming exhausted and driving up the cost.

World competition from the developing countires has begun to affect the price of prime energy. The Middle East fully knows that users will pay far more than the cost of extraction to supplement their petroleum supplies.

In the United States, conservationists have fought offshore exploration, new pipelines, refineries, and nuclear plants, making efforts to meet ever-increasing demands for energy more difficult.

The cost of energy as a factor in calculation of the gross national product (GNP) had dropped steadily for decades, but in 1966 the trend reversed. The ratio of energy cost to GNP is not expected to decline in the near future.

An ever-increasing quantity of solid waste with a high BTU content is generated. Approximately half of this waste, by weight, is replaceable. If all of the residential and commercial waste disposed of yearly were incinerated, the heat recovered would amount to less than 3 percent of the nation's total prime energy needs, or less than 10 percent of the heat energy required for heating and cooling residential and commercial buildings.

Conversion of solid waste certainly is not the whole answer to the rising cost of energy and the apparent depletion of certain fossil fuel supplies. It is, however, an economical and practical method for disposing of solid waste — swapping "cash for trash" by utilizing the heat energy for the many different applications once served by some of the low-priced nonrenewable fossil fuels.

Solid waste can be a blessing in disguise. An evaluation of the markets or applications for the heat energy available in solid waste follows, with emphasis on:

1. initial investment
2. fixed charges
3. operating costs
4. load factors by markets and their impact on production costs
5. the type and temperature levels of the fluids used for conveying the heat

There are numerous markets or uses for the heat recovered from solid waste. Some that have proved practical are:

1. process needs in manufacturing facilities
2. industrial parks in which process and/or factory comfort cooling or heating is required
3. metropolitan urban areas where new as well as existing offices, stores, banks, hotels, etc., require year-round climate control
4. college and university campuses with existing central heating and cooling plants serving academic, dormitory, administration, student union, athletic, library, and other areas on campus
5. medical centers with laboratory, research, surgery, patient, and administrative areas
6. airport complexes with central heating and cooling facilities for environmental control for each building
7. shopping centers where sales areas, malls, and miscellaneous spaces are supplied with coolant and medium-temperature and humidity
8. apartment complexes with high-rise facilities as part of an urban development program
9. power generation where solid waste heat energy may supplement the prime energy for base load or peak shaving usage.

Each of these applications or markets requires a different amount of coolant and heat per unit of demand. In other words, each will have a different load factor. Load factor as used here is the ratio of yearly unit steam sales to the potential yearly production. Expressed more succinctly, load factor is the steam sales and/or equivalent ton-hours per year in pounds divided by the steam produced in pounds. In-plant usage, line losses, and miscellaneous losses, therefore, are excluded.

Total steam production may be that produced only by the solid waste and supplementary fuel. These load factors are ratios of steam sold to that produced by solid waste plus supplementary fuel. Potential yearly production must allow for scheduled and unscheduled outages of the incineration equipment.

The load factor range for each of these applications is shown in Table 11-3.

Table 11-3 Load Factors

Applications (markets)	Load factor range
1. Manufacturing facility	
(process use only)	0.70-0.80
2. Industrial park	
(heating only)	0.25-0.35
(heating and cooling)	0.50-0.60
3. Metro urban area	
(heating and cooling)	0.60-0.70
4. University campus	
(heating and cooling)	0.40-0.50
5. Medical center	
(heating and cooling)	0.47-0.57
6. Airport complex	
(heating and cooling)	0.55-0.65
7. Shopping center	
(heating and cooling)	0.35-0.45
8. Apartment complex	
(heating and cooling)	0.37-0.47

The first item in Table 11-3, the manufacturing process that requires a fixed quantity of steam for each operating hour, will have the highest load factor of any of the markets analyzed — between .7 and .8. The demand is affected solely by the production process. It is not dependent upon the weather or outside conditions that affect transmission and, of course, ventilation air-heating.

The second market analyzed, an industrial park, has a lower load factor because only building heating is required and the consumption will be quite minimal. If heating and cooling of facilities in an industrial park are provided, the year-round need improves the load factor. The spring and fall consumption, however, will be low, as affected by the outside temperature. The load factor is on the order of .5 to .6; for heating only, the factor is .25 to .35.

The remaining markets in Table 11-3, in which the coolant and heating needs are influenced primarily by the weather, have substantially the same seasonal demands and yearly load factors. In some instances, the internal component of the coolant load has a noticeable impact on the consumption.

These load factors, going down the scale, vary from .4 to about .65. They cover such applications as urban area development programs in downtown areas in cities, university campuses, medical centers, airport complexes, shopping centers, and apartments.

The load factor has a significant impact on production costs, whether it be steam, medium or high temperature water, or coolant. It is axiomatic that the higher the load factor, the lower the production cost; therefore, applications such as manufacturing with a daily and year-round demand for heat energy will have the lowest production cost.

Initial investment for a solid waste incinerator facility with heat recovery and pollution control equipment will be affected by several different factors, specifically:

1. plant site, usually expressed in daily tons of solid waste
2. type of pollution control equipment
3. type of solid waste preparation and handling systems
4. type and quality of heat energy produced, together with type of fluids used for transfer from plant to point of usage
5. area of country (that is, Northeast, Southeast, West, etc.) and, of course, the economic climate prevailing

Some of these data can be incorporated in curves such as shown in Figures 11-1 and 11-2. This is expressed in daily tons of waste and steam output for the equivalent plant size. The coordinates are the initial investment in dollars per daily ton waste and the cost of production of energy, including the fixed charges. Included in these curves are the site, adaptation of the site with facilities, buildings, all hardware, storage areas for handling waste material, air pollution devices, all factory control devices, all stack pollution control devices, and, of course, the water wall type furnace, cranes for handling, mixing, etc., including the main heat exchanger, superheater, and economizer section. A moving grate assembly provides turnover and mixing to obtain more complete combustion.

The bottom curve, which is the overall cost of production, includes the fixed charges and operating costs for a facility of one particular load factor and is identified in the explanation at the top of the chart for a load factor of .5.

This includes auxiliary oil burners and forced and induced draft fan assemblies. The stack gas pollution control equipment is for both particulate and gaseous pollutants.

The building is of standard industrial type construction with solid waste storage pits large enough for at least three full days of operation.

Figure 11-1 Budget Cost For Small Size Incinerators With Heat Recovery

A. INITIAL INVESTMENT INCLUDES –

incinerator unit with waste heat recovery,
pollution control equipment, stack & breeching,
building with solid waste storage, piping, wiring,
controls, charging system on site labor, fees,
supplementary fuel facilities, financing costs
during construction.

B. OWNING & OPERATING COST INCLUDES –

fixed charges based on 20 year depreciation,
7% interest, taxes, insurance, operating personnel,
electricity, supplementary fuel, water, chemicals,
routine maintenance, service contracts and with
yearly load factor = 0.50

B-1. REDUCED OPERATING PERSONNEL

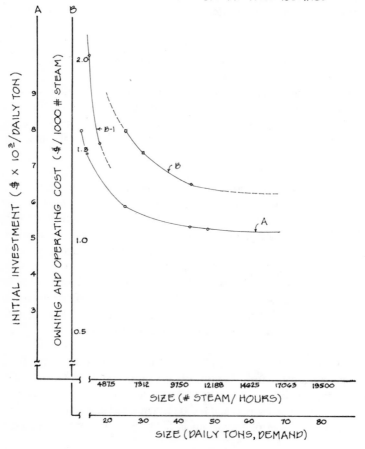

⊗ BASED ON 4500 BTU/# SOLID WASTE 65% CONVERSION EFFICIENCY

Figure 11-2 Budget Cost for Incinerators in Southeast, Adjusted to 1975 Level

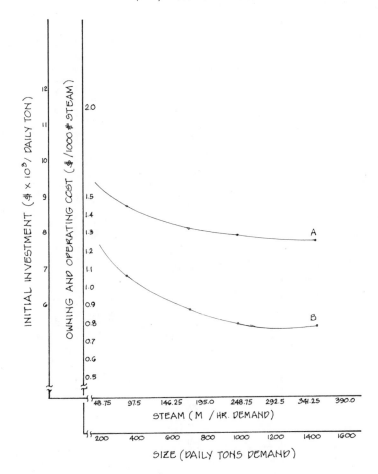

A. INITIAL INVESTMENT INCLUDES –

incinerator unit with waste heat recovery, pollution
control equipment, stack and breeching and building
with solid waste storage, piping, wiring controls,
charging system, on site labor, fees, supplementary
fuel facilities, financing cost during construction

B. OWNING AND OPERATING COST INCLUDES –

fixed charges based on 20 year depreciation, 7%
interest, taxes, insurance, operating personnel,
electricity, supplementary fuel, water, chemicals,
routine maintenance, service contracts and with
yearly load factor = 0.50

INITIAL INVESTMENT ($ × 10³/ DAILY TON)

OWNING AND OPERATING COST ($/1000 # STEAM)

STEAM (M / HR. DEMAND)

SIZE (DAILY TONS DEMAND)

⊗ BASED ON 4500 BTU/# SOLID WASTE 65% CONVERSION EFFICIENCY

Figure 11-3 Flow Schematic of a Typical Incinerator Plant

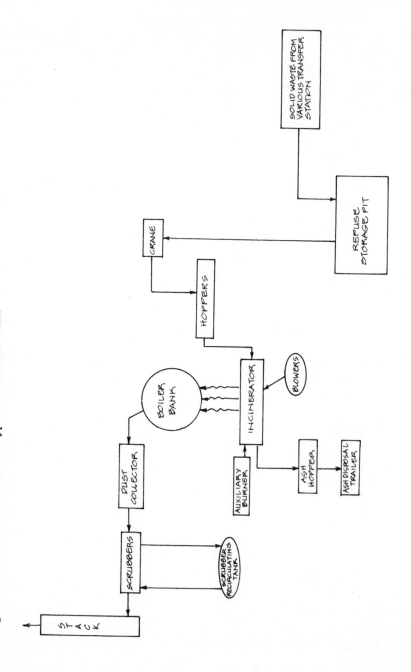

The piping, wiring, controls, on-site labor, and site preparation are included in the budget curve, along with fees for financing, design, legal counsel, and cost of money during the construction period.

Figure 11-3 shows a flow schematic through a typical incinerator plant. On the right is the solid waste from various transfer stations that is dumped into the storage pits. The storage capacity is equal to three days' operation (approximately 8,500 cubic yards). The waste can be mixed continuously by the crane operator before it is charged in the water-jacketed hoppers to serve the various incinerator units. Storage pits often have a capacity of three or four days' supply to carry them over long weekends.

Some of those are not designed to carry through for an extended outage when sanitary workers will not be on the job. Therefore, supplementary oil fuel (used oil available at filling stations and truck service stations) is provided. It is a very convenient way to dispose of the oil, which has a high BTU content. After settlement, it can be used readily in a plant like the one in Figure 11-3 without concern about emissions going up the stack.

Wastes move down the water-jacketed hopper throat onto a reciprocating type grate assembly for drying, ignition, combustion, and burnout, Residue and noncombustible materials are discharged into the ash disposal trailer. Separated material could be used for building materials, building blocks, or other markets that might be developed in the future.

The scrubbing of gases having approximately 150° F wet bulb with systems having about 100 percent saturation efficiency will produce stack gas at about 150° F dew point. This system would have approximately 100 percent saturation, so the gas would be at about 150° F dew point. Therefore, vapor plumes could form during portions of the year when the secondary or induced air lowers the mixture to the condensation point.

The induced draft fans have an inlet bypass so that reheated outside air may be mixed with the saturated gas from the scrubbers. This will lower the stack gas mixture dew point and, with reheat, will raise the dry bulb sufficiently to eliminate the vapor plume or "little white cloud." In addition, the stack gas is discharged at a velocity of approximately 3,500 feet per minute. Tests show that if any vapor plume or white cloud does form, it will be at some distance from the stack because of the elevated dry bulb and the depressed dew point.

All plants are required to use air pollution abatement equipment that will result in emission levels lower than those required by state and federal governments.

The emission levels will be considerably lower than those produced by the many small plants that the incinerator plant replaces. It is estimated that such a facility reduces annual particulate emission by a factor of four to five, even though the plant burns solid waste continuously and the many small plants it replaces burn fossil fuel only during the heating season.

Wet scrubbers are used so that gaseous pollutants, as well as particulates, can be removed. Bleed water from the scrubbers goes to the ash quench tank and is lost by evaporation and ash drag-out. Therefore, there is no net water effluent from the scrubbers and ash quench tank. Bleed from the cooling towers, if cooling has been provided, is used for makeup to the scrubber recirculation system. Boiler blow-down and plant cleanup water is fed to the sanitary sewer.

The cost of the chilling plant must be added to the total cost of the plant if the system is to provide coolant. Distribution system costs are not included in the curve in Figure 11-2. That must be added after determining whether steam or water will be used for heating.

When the amount of the initial investment for the basic plant has been derived from Figure 11-2 and distribution system(s) type and length have been established, these budget costs may be added to the basic plant cost and a total or new budget cost per ton determined. This new cost may be used to calculate fixed charges.

The distribution system(s) costs will vary depending on the type of fluid used — steam or hot and chilled water for an industrial park, medical center, airport, urban development, etc.

If a representative length is assumed for distribution, namely a comparatively short run for steam alone to a manufacturing plant and an extensive "finger" type distribution system for the two commodities to serve an urban development project, and the types of piping and insulation, excavation, filling, and resurfacing as required, a budget cost for this portion of the project may be established and added to the base costs obtained from Figure 11-2.

The incremental costs for the distribution system and complementary changes in the incineration plant are shown in Tables 11-4 and 11-5. The data were gathered from feasibility studies prepared in the last few years.

Table 11-4 Initial Investment Multiplier
(For Use With Figures 11-1 and 11-2, Curve A

	Investment multiplier
A. Manufacturing (process use only)	
a-1. Steam at $15\#/in^2$	1.080
a-2. Steam at $150\#/in^2$	1.060
a-3. Steam at $265\#/in^2$ and 100° F superheat	1.050
a-4. Steam at $400\#/in^2$ and 150° F superheat	1.045
B. Industrial park	
b-1. Steam at $150\#/in^2$	1.060
b-2. Steam at $150\#/in^2$, coolant at 41° F	1.560
C. Metro urban complex	
c-1. Steam at $150\#/in^2$, coolant at 41° F	2.050
c-2. Water at 290° F, coolant at 41° F	2.000

Table 11-4 continued

D. University campus
d-1. Water at 290° F, coolant at 41° F 2.210
d-2. Steam at 150#/in², coolant at 41° F 2.260

E. Medical center
e-1. Steam at 150#/in², coolant at 41° F 2.170

F. Airport complex
f-1. Water at 290° F, coolant at 41° F 2.130

G. Shopping center
g-1. Water at 290° F, coolant at 41° F 2.120

H. Apartment complex
h-1. Water at 290° F, coolant at 41° F 2.100

Table 11-5 Fixed Charge Multiplier
(For Use With Figures 11-1 and 11-2 Curve B

	Fixed charge multiplier
A. Manufacturing (process use only)	
a-1. Steam at 15#/in²	1.055
a-2. Steam at 150#/in²	1.041
a-3. Steam at 265#/in² and 100° F superheat	1.034
a-4. Steam at 400#/in² and 150° F superheat	1.034
B. Industrial park	
b-1. Steam at 150#/in²	1.041
b-2. Steam at 150#/in², coolant at 41° F	1.381
C. Metro urban complex	
c-1. Steam at 150#/in², coolant at 41° F	1.714
c-2. Water at 290° F, coolant at 41° F	1.680
D. University campus	
d-1. Water at 290° F, coolant at 41° F	1.823
d-2. Steam at 150#/in², coolant at 41° F	1.857
E. Medical center	
e-1. Steam at 150#/in², coolant at 41° F	1.796
F. Airport complex	
f-1. Water at 290° F, coolant at 41° F	1.768
G. Shopping center	
g-1. Water at 290° F, coolant at 41° F	1.762
H. Apartment complex	
h-1. Water at 290° F, coolant at 41° F	1.748

The cost additions are expressed as multipliers on the base cost of a conventional incinerator plant. As an example, to provide 150-pound steam to a manufacturer as shown by line a-2, the base cost will be increased 6 percent. This presumes a close-coupled arrangement with the plant on the manufacturing site and steam and condensate routed for a minimum of interference.

Should both coolant and steam be provided to several manufacturers in an industrial park (line b-2), the base cost will increase to about 56 percent. In this instance, a chilling plant is required. The facility, including two distribution networks with laterals, will be located on the complex. The result will be an approximate 50 percent increase over the basic cost.

The metro urban complex, as shown by market C, university campus market D, and the remaining markets analyzed require heating either in the form of medium temperature water or steam and coolant at about 40° F.

The increase in initial investment over that for the base plant is about 100 percent. This includes the chilling plant with auxiliaries such as heat rejection equipment, low head and primary pumping equipment, piping, service valves, etc., and the building. The incinerator facility would be remote and the services might be in the street, so an additional premium would be added for the distribution systems.

Total initial investment, thus determined, can be used for budgeting purposes. Fixed charges also may be ascertained.

Curve B in Figure 11-1 includes owning and operating cost data for the basic plant when operated for one load factor only.

The owning and operating cost correction factors for the different applications are shown in Figure 11-2. This multiplier when used with curve B will provide, for budgetary purposes, the cost of steam production when the plant load factor is 0.50. As an example, assume owning and operating costs for an incinerator plant serving a metro urban complex are required. This figure must include increased fixed charges for plant modification and for the distribution systems over and above those indicated by curve B. With a load factor of 0.50 and with steam at 150 pounds for heating and coolant at 41° F for cooling and dehumidifying, the multiplier for c-1 is 1.714. In other words, the total is 71.4 percent more than indicated by curve B. This presumes the load factor is 0.50.

Figure 11-1 indicated the load factor range for different applications. Corrected production costs due to deviation from the base load factor of 0.50 applicable to the market or application under consideration may be obtained from Table 11-6.

Table 11-6 Load Factor Multiplier
(For Use with Figures 11-1 and 11-2)

Application (market)	Load factor multiplier
A. Manufacturing (process use only)	0.816
B. Industrial park	
(Heating only)	1.445
(Heating and cooling)	0.936
C. Metro urban complex	0.843
D. University complex	1.081
E. Medical center	0.976
F. Airport complex	0.883
G. Shopping center	1.187
H. Apartment complex	1.141

Steam Service Rate

Demand charge

First	500 lbs./hr. of demand or any portion thereof at $100/mo.
Next	1,000 lbs./hr. of demand at $130 per month/1,000 lbs/hr.
Next	1,500 lbs./hr. of demand at $130 per month/1,000 lbs./hr.
Next	2,000 lbs./hr. of demand at $110 per month/1,000 lbs./hr.
Next	5,000 lbs./hr. of demand at $105 per month/1,000 lbs./hr.

All over 10,000 lbs./hr. of demand at $100 per month/1,000 lbs./hr.

Commodity charge

First	200,000 lbs./mo. at $1.50/1,000 lbs.
Next	300,000 lbs./mo. at $1.25/1,000 lbs.
Next	500,000 lbs./mo. at $1.00/1,000 lbs.
Next	1,000,000 lbs./mo. at $1.00/1,000 lbs.

All over 2,000,000 lbs./mo. at $0.90/1,000 lbs.

The minimum monthly charge would be the demand charge. A pound of steam is equivalent to one pound of water evaporated into, condensed from, steam.

This table indicates that for a metro urban complex C, a multiplier of 0.843 should be used to correct from the basic load factor of 0.50 to 0.65, in curve B in

Figure 11-2, the average for such applications. The net correction, therefore, is 1.714 × 0.843 = 1.445. These multipliers take into account the relative impact of owning as well as operating costs.

Generally speaking, the fixed charges for the incinerator facility with heat recovery, proper pollution control for both gas and liquid effluent, chilling plant, distribution system, etc., will be approximately 65 to 68 percent of the overall owning and operating costs. Of the remaining 32 to 35 percent of owning and operating costs, about 18 to 20 percent are dependent on plant output. This segment of costs includes operating personnel, management, and general administrative expense. Other operating costs such as electricity, water, chemicals, supplementary fuel, and routine maintenance and service vary with steam production. This segment is approximately 45 percent of operating costs or about 15 percent of owning and operating expense. As a result, the higher the load factor and yearly production, the lower the production costs. Expressed in another manner, if the system load factor were increased from the base factor of 0.50 to 0.60 (an advance of 20 percent), the total costs would grow only about 3 percent. In addition, if there were a 20 percent increase in gross income with only a 3 percent rise in owning and operating costs, the unit selling price could be reduced, if it were a "not-for-profit" operation, or the profit picture would be enhanced if it was a profit-making corporation.

As an example of the foregoing, assume that budget costs are required for initial investment, fixed charges, owning and operating costs, and production costs for a 1,000-daily-ton capacity solid waste plant with chilling plant sized for a heating-to-cooling demand ratio of 1.0 and serving an urban area with new and existing offices, stores, banks, hotels, etc. Figure 11-2 indicates that the load factor for metropolitan areas will vary between 0.60 and 0.70.

Figure 11-2, curve A, indicates an initial investment for a 1,000 daily-ton basic plant is $8,000 per ton, or $8 million. Included in this figure are costs as listed in Figure 11-2, item A.

Table 11-3 shows the multiplier for the addition of a distribution system and a chilling plant. For addition to market C, metro urban complex (using steam at 150 pounds per square inch for heating and 41° F coolant, the indicated correction factor is 2.050. Putting all these data together, the initial investment would become about $16.4 million for a 1,000-daily-ton plant.

In Figure 11-2, curve B indicates the overall owning and operation costs for a base plant when operated at a load factor of 0.50. From curve B, the owning and operating costs or production costs are $0.81 per 1,000 pounds of steam.

Table 11-5 shows adjustment in owning and operating costs (production costs) because of the additional expense of a distribution system and chilling plant. This correction multiplier for C, metro urban complex, with steam for heating and water for the coolant is 1.714; therefore, production cost is $0.81 × 1.714 = $1.39 per 1,000 pounds of steam.

Table 11-6 shows correction due to the metro urban complex load factor of 0.60-0.70 (use 0.65) instead of the 0.50 that is the base for curve B. This correction factor is 0.843; therefore, production costs are $1.39 × 0.843 = $1.17 per 1,000 pounds of steam.

The result of the above calculations is:

1. In-place cost for plant with distribution system is $16 million.
2. Production cost for services to the metro urban complex having a load factor of 0.65 is $1.17 per 1,000 pounds of steam.

Information such as this may be used during the preliminary stages of discussion and before the feasibility study is complete.

Should there be interest in return on investment at this preliminary stage, it is approximately 5 percent based on sales at $1.75 per 1,000 pounds of steam, an initial investment of $16,400 per ton capacity, and conversion efficiencies that provide 6,000 pounds of steam per ton (250 pounds of steam per hour per ton). However, if the municipality pays a nominal amount per ton for incineration in lieu of landfill ($3 per ton), the return on investment will be approximately 11 percent.

In time, economical techniques will be developed for recycling certain components and new markets will emerge that will provide additional income. Supplementary income for paving base or building block can be obtained from the residue after separation.

In summary, there will be an ever-increasing amount of solid waste available for conversion into energy. Disposition of the waste material will utilize the most economical techniques consistent with air, water, and land pollution constraints. Conventional incineration and pyrolysis — processes in which the energy may be recovered and put to an economical and useful purpose — are disposal methods that appear to have merit.

Engineers are called upon frequently to evaluate the economics of the many different disposal methods for solid waste. After a comprehensive study, they recommend to the municipality the best solution to the disposal problem. Some sections of the country do not have sufficient supplies of certain low-cost convenience fossil fuels to serve the needs of all industries in their area. The heat energy in solid waste may be a partial solution to this problem.

NUCLEAR POWER

Several types of nuclear reactors are available for production of power. They present numerous advantages and disadvantages that a management representative must weigh in selecting a system for modification of a plant.

Although the energy of fission appears as kinetic energy (85 percent) and photon energy of rays, there is no practical method of converting this energy directly into useful work on a large scale. Therefore, a nuclear reactor must be treated as a heat source that differs from a chemical heat source in that no oxygen is required and the heat does not have to be removed from gaseous combustion products that possess poor heat-transfer properties. Problems involved in use of heat-transfer systems limit their use to certain types of operations for turning nuclear material into useable energy.

The simplest possible cycle for heat-transfer devices is the Brayton-cycle, but for nuclear use, it presents problems of containment, contamination, high temperature, and high pressure. The Carnot-cycle efficiency is the maximum theoretically attainable for any heat engine, but this cycle has not been considered for a nuclear reactor because of the difficulty in transferring large amounts of heat isothermally. The Otto-cycle with its constant-volume heat addition, isentropic expansion, and constant-volume heat rejection, might be applied using either a solid or a gaseous fuel. However, little work has been done to test its feasibility.

Processes in the Rankine vapor cycle are effective for handling power in large amounts and therefore are used for central-station generation even though the efficiency of the plant is lower than for internal-combustion engine cycles. The Rankine cycle appears to be best suited for nuclear power because (1) the maximum practicable operating temperature for a reactor corresponds to or is lower than the temperature used in the conventional Rankine vapor cycle, (2) the problem of containment of radioactivity is better solved by this cycle than by other cycles, and (3) reactors are most economical in large sizes, as are Rankine engines.

Three variations of the Rankine cycle are being used. The direct cycle is the least expensive and thermodynamically the most desirable. Its disadvantages are that the boiling process does not permit the high power densities that are attainable with liquid cooling and that radioactive steam is carried into the turbine and condenser. The indirect cycle gives greater power density and eliminates radioactivity in the turbine, although it produces steam at lower pressures and temperatures than those considered most efficient for modern turbines. Both concepts are highly developed and available commercially. Plants utilizing sodium or other liquid metal as the heat-exchange medium use an intermediate-link system to isolate the water system from radioactive sodium.

Essentially, there are five main components in a reactor:

(1) atomic fuel, which must be sealed as protection against oxidation
(2) a moderator to slow down the neutrons so the chain reaction can continue efficiently
(3) a control mechanism to maintain the chain reaction at desired levels

(4) a coolant to maintain temperatures below the melting point of materials and, in the case of power reactors, to convey the heat to a steam generator

(5) a shield to contain radio-active by-products

The arrangement of moderator and fuel provides one basis for classification. In a heterogeneous or solid fuel reactor, the fuel is mixed in a regular pattern within the moderator. In a homogeneous reactor, the fuel and moderator, whether aqueous, liquid salts, or metals, are mixed intimately in the form of a solution.

A second classification is coolants, which include light water, heavy water, liquid metal, gas, and organic liquids. A further classification is by the speed or energy of the neutrons that cause fission. Neutrons of about 0.025 electron volts (ev) are called slow or thermal; neutrons from 1 to 1,000 ev are known as intermediate, and neutrons with energies above 1,000 ev are called fast. A fast reactor does not use a moderator.

Finally, reactors may be classified according to application: research reactors, designed to provide neutrons and rays for physical research and radioisotope manufacture; materials-testing reactors, generally classified as research although they run at higher power to produce a very high neutron flux; production reactors, which manufacture fissionable material by conversion of nonfissionable material, and power reactors for generating electricity to provide process heat and to propel ships and other vehicles.

Research and test reactors are a highly useful source of neutrons for a variety of experiments in the study of neutron properties, radiation damage, detecting and counting of rays, critical behavior, energy of fission, distribution of fission fragments, and related technology.

Radioactive isotopes, obtained by irradiation of samples with neutrons in a reactor, are used for tracing the progress of foods and minerals in plants and animals; studying blood processes, gland functions, and animal metabolism; investigating the properties of various elements and compounds, and measuring and detecting flaws in industrial processes.

A research reactor utilizing natural uranium has a large critical mass of U^{235} and therefore has a low neutron flux for a given power rating. It also is physically large, providing ample working areas around the reactor. Reactors using enriched uranium have a small critical mass that provides a high flux. All research reactors operate at low power and are relatively safe because they are designed to handle little excess reactivity and have a strong negative temperature coefficient.

When constructing a dual-purpose (production-power) reactor, the Atomic Energy Commission (AEC) revealed that the reactor consisted of a large block of graphite pierced horizontally by zirconium-alloy tubes containing the fuel elements and providing passage for coolant flow. This matrix is enveloped by thermal and biological shields. A primary cooling system, where it is utilized to generate steam for use in a turbine generator.

Power Reactors

Conceptually, a great variety of reactor types is possible. Although more than a score of combinations have gone through various stages of development, guidelines for selecting the maximum design have not been established firmly. In the United States, development of water-cooled reactors has outdistanced that of other types, while in England gas-cooled reactors hold the lead. It is not clear which, if either, of these types eventually will prove to be the better power producer. There are indications that to meet future demands for energy it will be necessary to obtain more efficient utilization of fuel than now appears possible in converter (burner) reactors.

Pressurized-Water Reactors

Properties of water and steam that led to their predominance as general purpose heat-transfer mediums also have caused their widespread application as reactor coolants. A major disadvantage of water is its relatively high vapor pressure. However, this can be overcome in part by allowing boiling in the reactor. Thermal efficiencies up to 36 percent are possible.

The use of water as a coolant and moderator is based on well-developed technology that indicates the ultimate size (or capacity) will be dictated primarily by heat-transfer requirements. The average heat flux q/A is around 300,000 BTU per hour per square foot, with a maximum of approximately 600,000. The resonance absorption requires enriched fuel. However, the cost of enrichment is justified economically because the increased power density reduces inventory charges. To provide sufficient neutron moderation, the H_2O-U^{235} volume ratio is kept slightly above 2:1.

The use of oxide fuel minimizes corrosion. Design problems are understood reasonably well, although costly structural provisions must be made to load and unload fuel because of the high pressures. Fuel enrichment usually runs between 1.5 and 3.0 percent, depending on the alloy used for corrosion resistance. Because of the strong moderating property of H_2O, control rods must be placed on about 1-foot centers. To prevent the large number of these rods from interfacing with the unloading of a large power reactor, they often are operated from the bottom.

For utilization of natural uranium as a fuel, D_2O (heavy water, composed of deuterium and oxygen) is substituted for H_2O as the coolant moderator. The reactor must be unloaded more frequently because the specific power is greater. The flux is higher, approximately 5×10^{13} for 1,000 mw, and there is a problem in overriding xenon. The success of this type depends to a large extent on the availability and cost of D_2O.

Boiling-water reactors have a simpler design and can utilize relatively thin-walled vessels and pipes because they operate at moderate pressures compared with pressurized-water reactors. Fuel temperatures are only slightly higher than steam temperatures, and there is an inherent safety factor because the steam-void volume increases on a transient power increase. Some of the problems caused by radioactivity carryover, such as maintenance on the turbine and condenser and the prevention of radioactive leakage, are offset by the cost of the boiler in pressurized-water reactors.

The limitation on the power density imposed by exit voids (i.e. the vapor volume of the exiting steam-water mixture from the reactor core) is a disadvantage in that low-power density contributes to high fuel-inventory charges. Load change must be accomplished by steam bypass control or rods since adjustment of the turbine throttle and consequent reduction in steam flow will cause an increase in pressure in the reactor, which in turn will collapse the steam bubbles and increase reactivity. The net decomposition of the water into O and H is much greater than in a pressurized-water reactor.

Gas-cooled reactors offer low fuel and operating costs because of their ability to use natural uranium as fuel. This advantage is offset by higher capital cost resulting from the larger facilities required by the low-power densities that prevail. The substitution of slightly enriched uranium will reduce capital expenditures, but it is not obvious how this change will affect economies associated with the fuel cycle itself.

Carbon dioxide generally is used as the coolant as it is nontoxic, only mildly radioactive, nonflammable, noncontaminating in case of leakage, and relatively inexpensive. Hydrogen gas has been ruled out as a coolant because of difficult technical problems, although its potential for superior performance is recognized. Helium gives indications of being an ideal coolant, but its limited availability and high cost make it unattractive in a power reactor. Full containment of the coolant is one of the major problems in a gas-cooled reactor because the natural warping and expansion of materials is aggravated by high operating temperatures.

Gas-cooled reactors offer a thermodynamic problem for which the solution depends largely on the methods used to remove the heat from the core. Both convection and conduction come into play, and these two phenomena differ in their nature and methods of handling. Extensive studies are being made of the heat-transfer and pressure-drop characteristics of finned cans and cartridges. Although the power consumption of the blowers required to circulate the cooling gas is an appreciable fraction of the power output, efficiencies approaching 30 percent have been achieved.

An upper limit on gas temperatures is created by the fact that the oxidation of graphite by CO_2 increases with a temperature rise. Limitations of size also are provided by the practical consideration of how large a pressure vessel can be constructed, stress-relieved, and tested in the field. Both prestressed concrete and all-steel construction have been used. From the point of view of reactor performance, there always will be an incentive to increase the physical size or pressure level of a given reactor, thus permitting the extraction of more power. Accordingly, the capabilities of pressure vessel fabricators play an important role in the final design of a gas-cooled system. The advantages of using liquid metal as a reactor coolant include its low vapor pressure, excellent heat-transfer characteristics, and the potential of producing steam conditions that are typical of modern generating station practice. Disadvantages stem from attendant chemical activity, neutron absorption, and induced radioactivity.

Choice of a liquid metal coolant is based mainly on nuclear properties and on engineering difficulties associated with melting temperatures and corrosion effects. Those of prime interest are sodium, sodium-potassium alloy, bismuth, lead, and lead-bismuth alloy. Separated Li^7 would be attractive were it not for its high cost. In general, ferrous alloys with low carbon content and high nickel and chromium content are preferred as materials containing liquid metals. Final choice is based on range. Solubility of oxide in liquid sodium increases rapidly with temperature (by a factor of 40 between 150° C and 500° C). Excessive oxide creates mass transport, plugging, and self-welding. Development of suitable pumps presents difficult problems. Considerable research has been done on pumping by electromagnetic means.

Liquid metal heat exchangers frequently are called upon to operate at much higher temperatures differences than are common in water types. Therefore, extra precautions are necessary to avoid thermal shock or thermal cycling. Precautions against leaks must be intense as most liquid metals react with both water and air.

Fast breeder reactors operate at extremely high power densities (as they are unmoderated) and are cooled by liquid metal. Their attractiveness stems from an ability to "breed," i.e., produce more fuel than they consume. Aside from savings in fuel cost, this characteristic is believed necessary to conserve the world's supply of energy.

Two conflicting definitions of a breeder reactor are in common use: (1) a reactor that uses a certain type of fuel; and (2) a nuclear reactor that produces the same species of material it consumes, regardless of the net gain or loss. The first definition has wider acceptance in the United States.

Advantages of operating in the fast-neutron spectrum include: (1) structural material for the reactor core can be selected without consideration of neutron absorption; (2) little excess reactivity is needed to compensate for fission-product buildup; and (3) reprocessed fuel in which fission-product removal may not be complete. The last attribute has the potential of reducing fuel costs substantially.

Disadvantages basically are similar to those of moderated liquid metal reactors, including the fact that sodium undergoes a violent exothermic chemical reaction on contact with water in the presence of air. The use of water and other similar coolants is ruled out because they also act as moderators. A critical problem is to develop a low-cost, high burnup fuel element.

Benefits from Nuclear Power

Whatever the speed with which nuclear fuels replace fossil fuels, certain benefits are considered likely to result. These include:

Equalizing Generating Costs

Fuel costs of stations fired by fossil fuel are influenced greatly by the cost of transportation. In some sections of the country, freight rates equal or exceed the purchase price of the fuel at source. As transportation is a minor factor in the delivered cost of nuclear fuel, expanding construction of nuclear plants will tend to equalize and lower generating costs. Thus nuclear techniques would enable utilities to maintain the trend toward lower generating costs (in constant dollars) that they have established in the past through continued improvements in the efficiency of steam plants.

Extending Energy Resources

On the basis of various estimates of fuel reserves and energy demands, the AEC calculated that if nuclear power or some other substitute were not introduced, the United States "would exhaust our readily available, low-cost supplies of fossil fuels in from 75 to 100 years and our presently visualized total supplies in from 150 to 200 years." Moreover, there is need to conserve fossil fuels for certain uses for which there is no practical present substitute, e.g., in transportation and as sources of industrial chemicals.

Nuclear power can be effective in extending energy resources because the energy potential in uranium and thorium reserves is considered to be many times that in remaining fossil fuel reserves. How completely nuclear fuel reserves are used depends on the course of future nuclear power development, in particular on the relative degree of emphasis placed on breeder reactors (which, as noted, produce more fissionable material than they consume) as opposed to the currently predominant converter reactors (which produce some fissionable material, but not as much as they consume).

If breeders are developed to an extent making possible utilization of all the fertile uranium and thorium, as well as the naturally fissionable U^{235} (0.7 percent of natural uranium), the AEC estimated the fission energy content of domestic nuclear resources as virtually limitless, over 300,000 Q, as opposed to estimates that the energy still available in fossil fuels is from 26 to 130 Q. (The energy unit Q equals 1 quadrillion BTU. It is estimated that between 3 and 4 Q will be consumed in the United States between 1963 and 2000.)

Other Benefits

The relatively negligible transportation costs of nuclear fuel can reduce greatly the existing interarea differential in power generation costs, making possible increased economic development in areas where power costs now are prohibitively high. If reductions in the cost of electricity brought about by nuclear power are substantial, they would have the effect of stimulating wider use of electricity. Because they depend less on fuel transportation systems, nuclear power systems would be less vulnerable to interruption during war, thus improving the county's defense posture. Finally, where coal fueled plants can present serious smoke pollution problems, nuclear plants are free of this difficulty.

NOTES

1. H. Tabor, "Selective Radiation: Wave Length Discrimination." Transcription of a conference of scientific uses of solar energy, vol. 2, part 1-A (University of Arizona Press, 1955) pp. 1-23.

2. J.T. Gier and R.V. Dunkle, "Selective Spectral Characteristics as an Important Factor in the Efficiency of Solar Energy Collectors." Transcription of a conference of scientific uses of solar energy, vol. 2, part 1-A (University of Arizona Press, 1955) pp. 41-56.

3. E.A. Christie, "Spectrally Selective Blacks for Solar Energy Collectors." Paper no. 7/81 from the I.S.E.S. Conference, Melbourne, Australia, 1970.

4. H.C. Hottel and B.B. Woertz, "The Performance of Flat-Plate Solar Collectors," *ASME* 64 (February 1942) pp. 91-105.

5. R. Erway and A. Whillier, "Thermal Resistance of the Tube-Plate Bond in Solar Heat Collectors," *Solar Energy*, vol. 8, no. 3 (July-September 1964) pp. 95-98.

6. E.A. Farber, "Solar Energy Conversion and Utilization," *Building Systems Design* (June 1972).

7. W.P. Green, "Utilization of Solar Energy for Air-Conditioning and Refrigeration in Florida." Master's thesis, College of Engineering, University of Florida, Gainesville, Florida (1936).

8. W.P. Teagan and S.L. Sargent, "A Solar Powered Combined Heating and Cooling System." Paper no. EH-94, presented at the I.S.E.S. Congress in Paris, France (1973).

9. G.O.G. Löf, "Unsteady State Heat Transfer Between Air and Loose Solids," *Ind. Eng. Chem.* 40 (1948) pp. 1061-1080.

10. F.S. Dubin, Testimony Before the Subcommittee on Energy, House of Science and Astronautics Commission, June 12, 1973 (published by Dubin-Minder-Bloom Association, New York.)

Chapter 12

Investment and Financing

HOLDING THE LINE ON ENERGY COSTS

The elimination of cost for wasted energy results in savings that can be counted on every year. However, financial planning of energy conservation measures is necessary to make sure the improvements result in payback on the long-term investment.

Specific actions and modifications of a building to reduce energy consumption depend on its size and complexity. Regardless of the type of structure, there will be a need to seek capital for certain modifications, and it will be necessary to make cost-effective use of this capital. Seeking funds and using them economically will pose varying problems to different plants, but there are some common energy finance and investment techniques that can be of assistance to any type of industry.

As the range of conservation measures and their respective costs is examined, it should be kept in mind that energy management is an evolutionary process. The initial focus should be on measures that appear feasible within the level of available staff and in-house expertise. A facility should not attempt to accomplish or implement all options immediately, but should move deliberately and carefully, choosing measures that have the greatest impact if they can be handled by existing organization. This step-by-step approach will ensure greater control over energy modifications and will assist in building the credibility needed to stimulate further funding of an energy management program.

SEEKING FUNDS OUTSIDE THE FACILITY

As a first step, potential sources of funds for an energy management program should be listed. Discretionary funds that may be available for general upkeep and maintenance, as well as sources of capital funds for major refit and systems-convert undertakings, should be investigated. The type of information that will be helpful for each source of funds includes:

- name of source and person in charge

- types of programs funded
- types of funds disbursed (loans, etc.)
- restrictions imposed
- time requirements
- format of request
- data required to support request

Questions on external sources probably can be answered by the agency involved. When possible, a few samples of previously approved proposals or budget requests should be obtained in order to have a better idea of what the lending agency considers a good request. Once a list of the sources of funds has been prepared, it should be reviewed. One should ask: Have all possible sources of funds been listed, including all public agencies. For example, the Department of Energy (DOE) is involved in cost-sharing contracts for demonstration of alternate energy projects. Professional and technical journals may provide information on other sources of funds.

Tools frequently used to perform financial analysis for energy investments are highlighted below. These tools are helpful for determining the range of energy investment alternatives available.

DEVELOPING LIFE CYCLE STUDY COSTS

The following example develops, on a step-by-step basis, an approach to determine alternate mechanical system life cycle study costs. This example shows what system criteria must be established as well as methods of establishing system first, operating, and maintenance costs.

Money evaluation (present value versus future worth; first cost versus operating cost savings) may be established according to various formulae. This dollar-worth information should be handled separately from the life cycle economics shown here. Decisions as to present value versus future worth of capital will influence the desired payback period, but also should be considered separately from the life cycle analysis. As increased demand is placed on energy resources, energy costs will increase. Inflation also causes energy costs to rise. Utility company demand charges are growing due to rising fuel costs. The result is rapidly advancing energy costs with no foreseeable end.

Life cycle studies will become increasingly important to determine the best of alternative mechanical system designs. The old method of selecting the system that has the lowest first cost is not a wise one. Systems are not built to last one year and be justified in that year. Most system designs anticipate a 10-year life or longer. The objective in system selections should be to evaluate the overall cost for the system life anticipated and make the selection based on that criterion.

The rapid increase in energy costs in many cases will justify the use of energy conservation apparatus at the expense of an increased system investment.

INDIRECT EVAPORATIVE COOLER

The example examines a low energy cooling system versus a conventional system. Indirect evaporative cooling equipment is used to precool 15,000 CFM of outdoor air. Compare the first cost, operating cost, maintenance cost, and replacement cost of a conventional (compression refrigeration) air-conditioning system with an indirect evaporative cooling system of the same capacity doing identical cooling work.

The term *EER (Energy Efficiency Ratio) below is an industry designation for energy conversion efficiency. The higher the *EER, the greater the efficiency.

Design Criteria

   a. design ambient conditions: 95° DB and 68° WB
   b. total air quantity: 15,000 CFM
   c. system operation: 1,800 equivalent full load operating hours
   d. electrical rate: $.035 per kwh plus $.01 per kwh (demand assessment); total: $.045/kwh
   e. water rate: $1.00 per 100 cubic feet of water (748 gal.)
   f. water use: .02 GPM/core/equivalent FL operating hour
   g. conventional system energy efficiency ratio (*EER): 7.5
   h. water, electrical rate, and replacement cost increase: 15% per year
   i. maintenance cost increase: 10% per year
   j. 10-year system life
   k. equipment replacement: indirect system $40 per ton - first year; conventional system $90 per ton - first year.

Equipment Selection

   a. two indirect evaporative coolers, five cores per unit operating in parallel at 1,500 CFM per core dryside and 900 CFM per core wetside.
   b. one common dryside fan: 15,000 CFM at .51" static pressure loss; indirect evaporative cooler dryside
   c. one common wetside fan: 9,000 CFM at .15" static pressure loss; indirect evaporative cooler wetside
   d. dryside fan (energy increase):
      (1.9 BHP) (900 w/BHP) .... 1,710 w
   e. wetside fan (energy increase):
      (1.5 BHP) (900 w/BHP) .... 1,350 w
   f. two water pumps at 150 watts .... 300 w
      Total energy used .... 3,360 w

For a conventional system, increase the size of the compressor, condenser, and related apparatus to accomplish the equivalent cooling capacity (18.9 tons) operating at an energy ration *EER of 7.5.

$$*EER = \frac{BTU}{watt} \quad \frac{(heat\ transferred)}{(electricity\ used)}$$

Performance
Indirect evaporative cooler low energy Equipment

a. entering conditions of 95° DB and 68° WB
b. leaving conditions of 81° DB
c. temperature drop = 14° F. (dryside)
d. (15,000 CFM) 14° TD) (1.08)=226,800 BTU/hr (18.9 tons)
e. indirect evaporative cooler: energy efficiency ratio =

$$\frac{226,800\ BTU/hr}{3,360\ watts} = *EER\ of\ 67.5$$

f. conventional system: assumed *EER of 7.5 or

$$\frac{226,800\ BTU/hr}{7.5} = 30,240\ watts$$

Annual Operating Cost

a. indirect evaporative cooler
   electric: (1,800 equivalent FL hours) × (3,360 watts).
   ($.045/kwh) = $272 first year
b. water: evaporation
   $$\frac{(10\ cores)\ (.02\ GPM/core)\ (1,800\ equivalent\ FL\ hrs.)\ (60\ min./hr.)\ ($1.00/100\ cu.\ ft.)}{748\ Gal./100\ Cu.\ Ft.}$$
   = $29 first year
c. assume water treatment required and bleed ratio equates evaporation = $29
   first year
   total water .... $58 first year
d. conventional system
   $$\frac{electric:\ (1,800\ equivalent\ FL\ hrs.)\ (30,240\ watts)\ ($.045/kwh)}{1,000\ watts/kw}$$
   = $2,449 first year

System First Cost

Indirect evaporative cooler

Quantity	System item	Estimated installed cost including all services
2	indirect evaporative cooler - 5 core each, includes pumps and rigging	$5,550
1	wetside fan and plenum	400
1	(assumed) increase in dryside fan motor size	150
1	wetside screened weather cowl (intake)	100
1	dryside screened weather cowl (intake)	100
1	(assumed) chemicator water treatment package	125
1	water piping and connection	125
	Total	$6,550

Conventional system

increase in condenser and compressor

$$capacity = 18.9 \text{ tons } \frac{(226,800 \text{ BTU/hr})}{(12,000 \text{ BTU/ton})}$$

Assume $500 per ton cost increase
(18.9 tons) ($500/ton) = $9,450

(Note: Assume the supply air and return air ductwork required would be approximately the same for either system.)

Maintenance Cost

Indirect evaporative cooler
(Note: labor rates applicable to general maintenance personnel)
Assume water treatment required

	First year cost
chemicals: ($8/month) (12 months)	$ 96
labor to add chemicals and inspect quarterly (4 visits/yr) (2 hr/visit) ($15/hr)	$120
drain and flush basin, replace sleeve-type water filters semi- annually: (2 visits/yr) (6 hr/visit) ($15/hr)	$180
Total first year cost	$396

Emergency maintenance (failure) allowance:	
labor: (10 hours) ($15) (*1.1)	$165 2nd year
material: ($75 (*1.1)	$ 83 2nd year
Total	$248 2nd year

* = First year emergency maintenance is covered by warranty, second
year maintenance to be performed at a 10% inflationary increase.

Conventional system
(Note: Labor rates applicable to refrigeration maintenance personnel.)
Assume air cooled condensers (heat rejection) with no water treatment re-
quired, but Bimonthly preventive maintenance required

(6/year (4 hours/visit) ($30.00/hour) ....$720 1st year

Emergency maintenance (failure) allowance:
labor - (10 hours) (30/hour) (*1.1) ....$330 2nd year
material - ($200) (*1.1) ....$220 2nd year
Total ....$550 2nd year

(Note: The failure potential, in dollars and probability, is higher with compression air-conditioning equipment. The indirect evaporative cooling system has a
small pump and wetside fan apparatus subject to potential failure. The conventional system has compressors, condensers, and fans subject to potential failure.)

Ancillary Equipment Costs

Ancillary items such as electrical wiring and circuit breakers, water makeup, drains, equipment support bases, and added building structural requirements must be considered. In most cases, the substantially greater electrical requirements (increased full load amperage, circuit breakers, wire sizing, etc.) of the conventional air-conditioning system will outweigh all other requirements of the ancillary services to facilitate the indirect evaporative cooler apparatus.

In this example, assuming 208V-60C-30 power, the reduction in electrical system requirements by using indirect evaporative cooling versus the conventional system would be approximately 70 full load amperes.

Replacement Fund

Indirect evaporative cooling:
 (18.9 tons) ($40/ton) .... $756 first year
Conventional system:
 (18.9 tons) ($90/ton .... $1701 first year

Summary

The foregoing data can be summarized as shown on the 10-year life "cost summary" chart. (Table 12-1)

Table 12-1 Cost Summary — 10-Year Life

	Conventional System						Indirect Evaporative Cooler System						Savings	
YEAR	First Cost	A	B	C	D	Total Cost	First Cost	A	B	C	D	Total Cost	Conv. System	Indirect Evaporative Cooler System
1	9450.00	720.00	2449.00	0	1701.00	14,320.00	6500.00	396.00	272.00	58.00	756.00	7,982.00	0	6,338.00
2		1342.00	2816.00	0	1956.00	6,114.00		684.00	313.00	67.00	869.00	1,933.00	0	4,181.00
3		1476.00	3239.00	0	2250.00	6,965.00		752.00	360.00	77.00	1000.00	2,189.00	0	4,776.00
4		1624.00	3725.00	0	2587.00	7,936.00		828.00	414.00	88.00	1150.00	2,480.00	0	5,456.00
5		1786.00	4283.00	0	2975.00	9,044.00		910.00	476.00	101.00	1322.00	2,809.00	0	6,235.00
6		1965.00	4926.00	0	3421.00	10,312.00		1001.00	547.00	117.00	1521.00	3,186.00	0	7,126.00
7		2161.00	5665.00	0	3935.00	11,761.00		1102.00	629.00	134.00	1749.00	3,614.00	0	8,147.00
8		2377.00	6514.00	0	4525.00	13,416.00		1211.00	724.00	154.00	2011.00	4,100.00	0	9,316.00
9		2615.00	7492.00	0	5203.00	15,310.00		1333.00	832.00	177.00	2313.00	4,655.00	0	10,655.00
10		2877.00	8616.00	0	5984.00	17,477.00		1466.00	957.00	204.00	2660.00	5,287.00	0	12,190.00
												TOTAL	0	74,420.00

A = Maintenance Cost   C = Water Cost
B = Electrical Cost   D = Replacement Fund

In cases where the first cost of the indirect evaporative cooling system installation is higher than the equivalent conventional air conditioning system, the life cycle study (cost summary) will indicate the payback period of time as well as the dollar savings through a 10-year period of system operation.

The example given here shows the following savings:

1st year	$ 6,338
2nd year	4,181
3rd year	4,776
4th year	5,456
5th year	6,235
6th year	7,126
7th year	8,147
8th year	9,316
9th year	10,655
10th year	12,190
Total	$74,420

CAPITAL RECOVERY FACTORS

The useful life component employed in life cycle costing should be no greater than the remaining useful life of the building or system involved. To forecast costs over an extended period, the effect of inflation on labor, materials, and energy must be anticipated and a reasonable rate of interest on borrowed funds must be assumed. If it can be shown on a purely financial basis that installation of equipment will result in savings, such options should be considered strongly. However, more than monetary factors have to be considered. It also is necessary to take into account the ease of maintenance, the possibility of fewer complaints, greater reliability, future availability of energy (at any price), and other considerations.

In utilizing life cycle techniques to select between two systems, simple formulae can be used to determine whether the additional initial costs of a more expensive system are merited in light of long-term savings.

Cost-Benefit Measures

The formulae used to analyze energy investments may be different at each plant. The two most common cost-benefit measures used to evaluate the relative merits of a given investment are payback period and return on investment (ROI). In some cases, alternate methods such as present-value analysis may be preferred. The type of calculation most applicable at a particular plant, or most appropriate for a specific fund source, should be determined. The choice of a method of financial analysis also depends not only on a company's accounting practice, but on the type of project.

Payback period =

$$\frac{\text{Amount of investment}}{\text{Net annual savings — Annual depreciation}}$$

ROI =

$$\frac{\text{Net annual savings — Annual depreciation}}{\text{Amount of investment}}$$

The initial cost, or amount of investment, is the full cost of installing the particular equipment or equipment modification. Net annual savings is the difference between the cost of the enregy that will be saved by the improvement minus all annual costs of owning and operating the equipment, such as interest and principal on borrowed funds, maintenance costs, insurance, and taxes if applicable. Annual depreciation can be calculated in several ways. Generally, a straight-line depreciation formula is used, wherein the cost of equipment is divided by its useful life. Straight-line is recommended for this calculation.

PAYBACK PERIOD

This method of financial analysis is best suited for projects with high annual savings relative to capital cost. Failure to consider cost of debt service is the most common error made when making a payback period analysis. It cannot be said, for example, that an initial capital investment of $50,000 that results in annual maintenance and operation savings of $10,000 has a payback period of five years. To do so would ignore the fact that interest must be paid on the loan, or that — if no loan is involved — the money otherwise would be earning interest. A simple graph (Table 12-2) is based on the ratio of C (initial cost of a system) over S (annual maintenance and operating cost savings). The horizontal axis shows the payback period in years. Each curve represents a different rate of interest. Thus, assuming an initial investment of $3,000 and annual operating and maintenance savings of $600 (which will be applied to repaying the loan), the result is a C/S ratio of 5 that, at 10 percent interest, results in a payback period of a little more than seven years.

If more accuracy is required, the following formula can be used:

$$n = \frac{\dfrac{S/rC}{S/rC - 1}}{\log (1 + r)}$$

where: C = capital cost
S = annual operating and maintenance savings
r = interest rate
n = number of years to achieve payback

Table 12-2 Capital Recovery Factor

	Interest Rate, r		
	10.0	**12.0**	**15.0**
5	0.26380	0.27741	0.29832
10	0.16275	0.17698	0.19925
15	0.13147	0.14682	0.17102
20	0.11746	0.13388	0.15976
25	0.11017	0.12750	0.15470
30	0.10608	0.12414	0.15230

RETURN ON INVESTMENT

This method is most useful for comparing energy investments with alternative investments. It also is helpful when borrowed funds are involved, in order to ensure that the annual savings will be sufficient to pay back the principal and interest as well as other annual costs. If the return on investment is significant, it generally is sufficient to undertake the Return on Investment (ROI) analysis for the first year, since increases in energy costs are likely to exceed those for maintenance and other annual expenses. Thus, an improvement can be expected over time.

Present Value

This method of analysis generally is preferred when the annual savings are low compared to the capital investment involved. Present value is simply the price an institution should be willing to pay now to realize annual savings in the future. If the capital cost of a project does not exceed that price ceiling, the project is worth doing. The reliability of the present value method depends heavily on the reliability of forecasts for inflation of fuel costs and other items. Since it is so difficult to forecast future energy costs with certainty, this method must be used with some caution when applied to energy management projects. In present value analysis, it also is not necessary to take depreciation into account, because the savings used in the calculation are limited to the useful life of the improvement. At the end of the useful life, a new present value determination can be made for replacement.

The following example presents the results of a sample return on investment calculation as they might appear on completion, as well as the way in which the return on investment data were derived.

Item	Cost
1. installation of heat exchanger in central chilled-water plant to permit shutdown of chillers when outdoor temperature is below 60°.	$60,000
2. gross annual savings	30,000
3. — annual maintenance cost	3,000
4. net annual savings	27,000
5. depreciation or replacement allowance	7,000
6. annual return on investment	33%

ROI =

$$\frac{\text{Net annual savings} - \text{Annual depreciation}}{\text{Amount of investment}}$$

or

$$33\% = \frac{\$27,000 \text{ (step 4)} - \$7,000 \text{ (step 5)}}{\$60,000 \text{ (step 1)}}$$

where step 4 = step 2 — step 3

Step 1 is analysis of the cost of the installation. The cost is the sum of the following six elements:

- feasibility study
- engineering plans and specifications
- equipment purchase
- equipment installation
- startup costs
- contingency costs

Step 2 is the determination of the annual energy savings expected from the project. If this is the first project of its kind at the institution, it will be necessary to rely on theoretical calculations by staff members of a consulting engineer. For example, if the project is designed to reduce the amount of fresh air taken into a building, the quantity of air reduction can be calculated for the year, as well as the amount of energy it would have taken to heat the saved volume of air to the required amount of degrees. The energy (BTUs) then can be converted into fuel consumed, taking into account the losses in the system. This quantity of fuel is multiplied by the unit cost of the fuel to produce the annual energy savings for fuel. To this amount should be added the cost of electricity saved as a result of less demand on the fan used to move the air.

Such theoretical calculations generally are reliable, but they can be confirmed by subsequent metering of typical installations. Again, it is recommended that an evolutionary, step-by-step approach be followed so that theoretical estimates can be confirmed and overcommitment to projects that will not prove to be cost-effective can be avoided.

Step 3 is determination of the annual costs of ownership and operation. This involves estimating or delineating the following elements:

- routine maintenance of equipment, including labor and parts
- allowance for major repairs, if warranted
- calibration and maintenance of controls
- annual interest payments on borrowed money
- annual principal payments on borrowed money
- taxes and insurance

Step 4 is the calculation of the net annual savings. This is done by subtracting the annual maintenance and ownership costs from the annual energy savings (step 2 minus step 3).

Step 5 is the determination of the annual depreciation or replacement costs. An institution may have its own formula for this calculation, but generally straight-line depreciation can be used, i.e., dividing the equipment installation cost by the years of useful life.

Step 6 (optional) involves the determination of the inflationary factors for maintenance, fuel costs, etc. Generally, this type of analysis is not necessary for fast-payback items.

The same type of procedure and analysis would be used for a payback calculation:

$$\frac{\$60,000 \text{ (step 1)}}{\$27,000 \text{ (step 4)} - \$7,000 \text{ (step 5)}} = 3 \text{ (years)}$$

To calculate energy savings to be obtained from any modification, savings should be counted only once. For example, if savings of fan horsepower as a result of a timeclock installation have been counted, those savings should not be counted again when it is proposed to connect the same fan to a computerized control system. In this instance, credit should be taken only for the additional savings that will result from a more flexible shutdown capacity.

It might help to calculate both return on investment and payback. In this way, the priority project would be the one that is better from both calculations.

Benefit/Cost Analysis

Benefit/cost analysis can be used in two ways: to determine whether it is worthwhile to replace an existing system with a new system, or to determine which of two or more alternative new systems provides the better cost benefit.

To determine whether it would be more advisable to retain existing equipment or obtain new, more efficient equipment, the following factors must be considered:

1. maintenance savings to be obtained by installing the new equipment
2. operating and energy cost savings to be obtained by savings from installing the new equipment
3. salvage value of old equipment
4. capital cost of new equipment including legal fees, design professional fees, installation charges, cost of equipment, etc.
5. costs of financing
6. changes in property taxes as a result of installing new equipment
7. income tax factors
8. changes in rental income as a result of installing new equipment

If it can be shown on a purely monetary basis that new equipment will result in savings, obviously it should be considered strongly. But more than monetary factors should be evaluated. The company also must take into account the ease of maintenance and possibly the ability to do away with certain positions, reduced number of tenant complaints, greater reliability, and so on.

When utilizing benefit/cost analysis to select between two systems, a simple formula can be used to determine whether the additional initial costs of the more expensive system are merited in light of long-term cost factors. The result is a benefit/cost ratio that, if it exceeds 1, indicates extra initial expenses will result in long-term savings, as shown in the example below.

Total first cost System A	$18,000	
Total first cost System B	20,000	
System B exceeds System A by		$2,000
Annual operating, maintenance, and energy cost, System A	$ 1,600	
Annual operating, maintenance and energy cost, System B	1,000	
System A exceeds System B by		$ 600

Assuming a 20-year useful life for each system and a 10 percent interest rate, the capital recovery factor (a factor that, multiplied by the total loan amount or total principal, yields the annual payment necessary to repay debt) can be computed from Table 12-1. In this case, it would be 0.11746. Given this, the amortization cost for additional capital investment of System B would be .11746 × $2,000, or $234.92. Therefore, the benefit/cost ratio for System B would be:

$$\frac{\text{annual savings}}{\text{amortization cost}} \quad \frac{\$600.00}{\$234.92} = 2.55$$

Because the benefit/cost ratio exceeds 1, System B, although initially more costly, should provide more long-term savings.

Initial Management Efforts

Initial management efforts required for the total energy management approach are not dissimilar from other initial organizational steps that must be undertaken when understanding a new venture.

First, management must determine what specifically is to be done to effect energy conservation. This includes allocation of resources (personnel, money, etc.), development of policies and procedures to be followed, assignment of specific accountable responsibilities, obtaining cooperation of operating employees and tenants, determining what changes need to be made to operating and maintenance procedures and schedules, noting which equipment is to be repaired, adjusted, or replaced, and so on.

Second, management must determine precisely what must be done to effect each change, in terms of meetins, hiring additional personnel, ordering materials, etc.

Third, the approximate amount of time involved to achieve each objective must be settled. For example, if signs must be printed, one week may have to be allowed for preparation and approval of copy and graphics, two weeks for printing, one week for distribution, and three or four days for posting.

Fourth, it must be decided when certain objectives must be completed. When this is done, of course, it automatically will determine when work on the objective should be started by using data obtained through the third step, above.

Fifth, master schedule must be prepared to show exactly what will be happening. Competent personnel should be assigned to the various functions. The system should be examined for possible conflicts, such as having someone scheduled to do something during vacation week, etc. These should be rescheduled to avoid conflicts.

Everything cannot be expected to run smoothly. There are bound to be problems, schedules not met, people failing to do what they were assigned to do, etc. By having a master schedule, however, adjustments can be made relatively easily. As time moves on, it will be possible to develop even more specific monthly schedules, most of which will contain routine procedures refined through experience.

Chapter 13

Management
Case Studies

CASE HISTORIES OF BUILDINGS IN CHICAGO AND NEW YORK CITY

Following are case histories providing details on buildings where energy conservation activities have been undertaken. Results of the energy management efforts are based on comparisons of consumption patterns of the same period in the previous year.

Each case shows the characteristics of the building, the actions or modifications instituted, the amount of money saved annually as a result of the changes, and the length of time required to realize a return on the investment.

Building Location:	Chicago, Illinois
Type:	Insurance offices
Building condition:	Modernized, kept up to date
Area:	556,000 gross square feet
Type of HVAC system:	1,400-ton central plus 300-ton packaged
Number of floors:	19
Built:	1939
Energy sources:	Electricity and purchased steam for cooling and heating.
Average occupancy:	95%
Hours of daily occupancy:	7 a.m. - 6 p.m.

The Owner/manager established energy conservation measures in November 1973. These measures consisted of, among other things:

- setting thermostats at 68° F during the heating season
- reducing steam to noncritical building areas
- reducing temperature of hot air decks in equipment rooms
- banning use of supplemental heaters
- removing obstructions from perimeter heating units

- serving cold lunch one day per week to reduce steam consumption
- reducing domestic hot water temperature
- reducing lighting levels by an average of 50 percent in noncritical areas (restroom, elevator, elevator lobbies, corridors, decorative lighting in lobby)
- turning off lights in areas not in use (conference rooms, supply rooms, vacant offices, unoccupied space)
- turning off exterior building lights
- encouraging employees to turn on lights only as needed an area at a time, to turn out lights when leaving a space for 20 minutes or more, and to turn off lights near the end of the day
- eliminating lights on Christmas tree ornamentation
- closing venetian blinds and drapes during cold, windy days to provide additional insulation
- discontinuing sterilization of drinking water
- installing more efficient oven in kitchen

The results of the program for the periods are as follows:

Electricity (kwh)			
Month	**1972/1973**	**1973/1974**	**% Savings**
December	573,656	477,000	16.8
January	644,028	506,000	26.1
February	567,000	474,000	16.4
March	595,440	506,000	15.0
		Average	18.575

Steam			
Month	**1972/1973**	**1973/1974**	**% Savings**
December	2,945	1,806	38.7
January	4,200	3,794	9.7
February	3,120	2,987	4.3
March	3,664	2,455	33.0
		Average	21.425

Building Location: New York, N.Y.
Type: Office, general
Building condition: modernized

Area:	574,000 gross square feet
Type of HVAC system	1,800-ton central plus 80-ton package
Number of floors:	17
Built:	1912 (addition 1932)
Energy sources:	Electricity, purchased steam, and natural gas for cooling and heating
Average occupancy:	95-98%
Hours of daily occupancy:	8:30 a.m. - 4:30 p.m.

The owner/manager established conservation measures in September 1973. These measures consisted of, among other things:

- setting building thermostats at 75° F during cooling season
- setting outside air intake dampers to minimum position
- lowering temperature of domestic hot water
- reducing lighting levels in corridors, wardrobes, washrooms, and unoccupied areas by delamping of fluorescent fixtures and relamping with "energy saver lamps"
- replacing certain fluorescent lamps in office areas with "energy saver lamps"
- modifying operating schedules
- reducing elevator service during off-peak hours
- changing cleaning hours from 10 p.m.-6:30 a.m. to 4 p.m.-12:30 a.m.
- reducing lighting levels during cleaning hours

The results of the program for the periods are as follows:

Electricity (kwh)			
Month	**1973**	**1974**	**% Savings**
January	1,012,000	996,000	1.6%
February	1,062,000	1,067,000	(.5%)
March	1,016,000	922,000	9.3%
April	980,000	924,000	5.7%
May	1,006,000	962,000	4.1%

Steam (MLBs)			
Month	**1973**	**1974**	**% Savings**
January	3747.9	3530.0	5.8 %
February	4273.8	4120.3	3.6 %
March	2728.9	2668.4	2.2 %
April	2232.9	1631.8	26.9 %
May	1679.6	1455.1	10.38%

REPORT TO IMPLEMENT ENERGY CONSERVATION MEASURES AT A MAJOR UNIVERSITY

Introduction

Electrical energy use history of the building was obtained from metering records. Metered data were not available for the steam used; however, annual steam usage was approximated by analysis of loads, weather conditions, and present system operation.

Energy conservation techniques were applied to the facility, and their effects were established by a mathematical investigation. This report briefly discusses recommended energy conservation measures, their respective implementation costs, and return on investment considerations. The energy conservation measures discussed are:

Item No. 1: Effect on energy consumption for unoccupied cycles of operation on several air handling systems

Item No. 2: Effect on energy consumption for reduced lighting and air flow for several air handlers (with unoccupied cycle of operation assumed implemented)

Item No. 3: Effect on energy consumption for conversion of constant volume reheat systems to variable volume reheat systems (with unoccupied cycle of operation assumed implemented)

Item No. 4: Effect on energy consumption for instituting the economizer cycle of operation for air handling units

The collective effect on energy consumption upon implementation of all of the system modifications also was examined. Reduction of peak lighting load (due to the removal of fluorescent tubes) affects variable air volume savings, so that collective energy savings cannot be added.

Table 13-1 Effect of System Modifications on Energy Consumption

Gross Area: 119,671 sq ft
Supply Air: 141,175 cfm
cfm/sq ft.: 1.18
Original B-sq. ft./yr: 458,368
Revised B/sq.ft./yr: 204.313

	Base Past Energy Use (1)	Item # 1 Unoccupied Cycle of Operation For Several Air Handlers	Item # 2 Reduced Lighting & Air Flow For Several Air Handlers (2)	Item # 3 Conversion Of The Constant Volume Reheat Systems To Variable Volume Reheat Systems (2)	Item # 4 Institute Economizer Cycle Of Operation For AH-2 & AH-3 (2)	Item # 5 Sum Effect Of All Modifications
Electrical Consumption (1000's kwh)	2,887	2,295	1,740	2,230	2,295	1,720
Equivalent Gas Consumption (MCP)	45,000	26,540	23,755	23,675	22,540	18,580
Cost of Energy (Dollars) (3)	114,420	78,560	64,420	73,495	72,440	57,130
% Dollar Reduction For Modification	— —	31	12	4	5	50
Savings in Energy Costs (Dollars)	— —	35,860	14,140	5,065	6,120	57,290
Estimated Implementation Cost (Dollars)	— —	19,500(4)	15,240	17,980	8,400	55,240(4)
Return On Investment (Months)		6	13	42	16	12

1) Electrical usage was determined from accurate metering records. Gas usage was estimated from analysis of load and weather conditions. 4% electrical line losses were added to the metered consumption.
2) Unoccupied cycle has been assumed implemented. Savings represent those beyond unoccupied cycle savings.
3) Costs of gas and electricity used were June, 1975, utility rates: 1.87 cents per kwh for electricity and $1.34 per mcf.
4) This cost can be reduced by $10,000 due to partial accomplishment through phase I.

Summary and Conclusions

The effects of the proposed conservation measures on the total consumption of energy are summarized in Table 13-1, which is the basis of the following discussion:

Item No. 1: Effect on energy consumption for unoccupied cycles of operation on several air handling systems

  a. The facility normally is used Monday through Friday between 8 a.m. and 11 p.m. Evening usage generally is confined to the auditorium or large lecture rooms; both areas are served by one air handler, AH-1.

  b. When completed, unoccupied operation for AH-1 will be as follows: Shutdown of the main fans (SF-1, RF-1), exhaust fans (EF-D, E, H), and circulating pumps will be accomplished through sequencing of a time clock. Shut down will occur at 11 p.m. Monday through Friday, and the system will restart at 7 a.m. on weekdays. A low limit thermostat will cycle the system upon demand during the winter season. A manual override will be located in the mechanical room to accommodate occasional weekend use of the auditorium and lecture rooms. It should be noted that AH-1 was inoperative this past summer. This practice may not continue in the future, if usage of the auditorium varies. Reported savings reflect a normal unoccupied cycle of operation; if AH-1 remains off in the summer, a greater saving will be realized.

  c. When completed, unoccupied operation for AH-2 and AH-3 also will be controlled by time clocks providing sequenced shutdown of the main fans (SF-2, RF-2, SF-3, RF-2), exhaust fans (EF-F, G, I, J), and circulating pumps. Shutdown schedule is as follows: System will be inoperative at 5 p.m. Monday through Friday and will come back on line at 7 a.m. weekdays. A low limit thermostat will restart air handling system(s) upon demand during the winter season.

  d. The lab area unit (AH-4) will be shut down during the same hours as those listed in (b) by the clock control. A manual override will be installed to be used in the event of necessary experimental work during usual unoccupied hours. Lab area supply fan, and EF-A, B, and C will be shut down during unoccupied operation.

  e. The two direct expansion units will be shut down at 11 p.m. Monday through Thursday, and 5 p.m. on Friday by time clock control. The units will restart at 7 a.m. weekdays. The shutdown schedule is coordinated with present computer schedule is coordinated with present computer usage. Variance in the future can be handled by adjusting the time clock control.

 f. The primary pumps will be shut down by relays when all the air handlers go off. The pumps will sequence on when any air handler comes on line for either low limit or occupied operation.

 g. Table 13-1 shows a total energy saving of $35,860 at an estimated cost of $19,500 for a return on investment in 6 months. Much of this will be accomplished by the Central Campus Environmental Control Center Contract, reducing the actual funds required for implementation to $9,500.

Item No. 2: Effect on energy consumption for reduced lighting and volume levels for several air handlers

 a. A total of 4,342 tubes will be removed from the building. This reduction in lighting level allows a smaller air flow volume due to reduced cooling load. Savings are reflected in cooling costs, fan operating costs, and lighting costs.

 b. Reduced lighting level and air flow volume will be accomplished at a total cost of $15,240 and will save an additional $14,140 beyond unoccupied cycle savings, yielding a return on investment in 13 months.

Item No. 3: Effect on energy consumption for conversion of constant volume reheat systems to variable volume reheat systems

 a. Many areas of the facility operate at maximum occupancy for only a few hours a week, so supply air volume can be varied at other times to reflect the load. As an example, the large first floor auditorium (Room 131) is used as a lecture room for only a few hours per day and for assembly a few evenings a week. At other times, the room generally is unoccupied and the load is small. In the latter situation, the zones can supply a smaller air flow, leading to reduced cooling load, reduced reheat load, and reduced fan load for considerable energy savings.

 b. Motor-operated volume dampers will be installed for AH-1 on terminal units controlled with the reheat coil valves to vary supply air volume. Minimum air quantity requirements will be maintained so that proper ventilation is ensured.

 c. Large classrooms on the second, third, and fourth floors typically are supplied through four mixing boxes; two are located near a room's exterior exposure, the other two serve the interior portion. A sheet metal blank-off will be installed on the hot deck connection of the interior boxes, allowing only variable volume cooling through these boxes. This conversion to partial variable volume will reduce false loading of the systems and permit reductions in fan loads to obtain savings in both electrical and steam consumption.

d. Table 13-1 shows the systems can be converted in part to variable volume for an estimated implementation cost of $17,980. The measure provides annual energy savings of $5,065, for a return on investment in 42 months.

Item No. 4: Effect on energy consumption for instituting the economizer cycle of operation for AH-2 and AH-3

a. In the past, building pressure problems have not allowed successful economizer cycle of operation. Changeover to minimum outside air has been at a point lower than optimum, so that mechanical cooling was required even in moderately cold weather.

b. Outside air, return, and relief dampers will be replaced with low leakage dampers. The present birdscreen ($\frac{1}{4}''$ mesh) on the outside air intake of AH-3 will be replaced with $\frac{3}{4}''$ hardware cloth. The expanded metal screen on the relief dampers of AH-2 also will be removed. One of the three sections of the outside air intake of AH-2 will be blanked off to reduce outside air intake area to its proper size. These measures along with reduced air volume should help resolve the pressure problems and allow successful economizer cycle operation, with changeover point at 70° F.

c. The table also shows a successful economizer operation will save $6,120 in annual operating costs at an estimated implementation cost of $8,400, for a return on investment in 16 months.

Detailed Cost Estimates for Implementing Conservation Measures

Modification 1 – Unoccupied

AH-1

Shut down SF-1, RF-1, EF-D, EF-E, circulating pumps

Time clock	$ 400	
Controls and wiring	1,200	
Override control	100	
2 Low-limit thermostats	500	
Total		$ 2,200

AH-2 & AH-3

Shut down SF-2, RF-2, EF-1, EF-J, circulating pumps, ST-3, RF-3, EF-F, circulating pumps

2 Time clocks	800	
Controls and wiring	2,400	
4 Low-limit thermostats	1,000	
Total		4,200

AH-4 (Lab Area)

Shut down SF-4, EF-A, B, C, circulating pumps

Time clock	400	
Controls and wiring	1,400	
Override control	100	
Total		1,900

DX-5 and DX-6 (Computer Area)

Time clock	400	
Controls and wiring	1,000	
Override control	100	
Total		1,500

Chilled and Heating Hot Water Systems

Shut down at night after AH systems are off

Controls and wiring	3,500	
4 Ventilation overrides @ $1,000	4,000	
Engineering and contingencies	2,200	
Total		9,700
Grand Total		19,500

Modification 2 – Reduced lighting and air flow

4,342 tubes removed @ $1/tube	4,342	
New fan drives		
3 S.A. fans @ $800	2,400	
3 R.A. fans @ $700	2,100	
Total		4,500

Rebalance mixing boxes & zone dampers

10 Zone dampers @ $90	900	
150 Mixing boxes @ $30	4,500	
Engineering and contingencies	998	
Total		6,398
Grand Total		15,240

Modification 3

New fan drives*		
3 S.A. fans @ $800	2,400	
3 R.A. fans @ $700	2,100	
Blank-off plates		
66 Mixing boxes @ $30	1,980	
10 Motorized volume dampers @ $500	5,000	
Motor operators and controls for inlet vanes		
3 S.A. and R.A. fans	4,500	
Engineering and contingencies	2,000	
Total		6,500
Grand total		17,980

Modification 4 — Economizer cycle

Install new low leakage dampers on AH-2 system

O.A. damper, return damper, relief damper	1,000	
Labor to remove old dampers and install new dampers; birdscreen blank-off	1,000	
Provide 3 additional damper operators and install baffle on return fan discharge side	500	
Engineering and contingencies	1,000	
Total		3,500

Install new low leakage dampers on AH-3 system

O.A. damper, return damper, relief damper	1,500	

Labor to remove old dampers
  and install new dampers to
  include replacement of birdscreen
	1,200	

Relocate outside sensor	200	
3-damper operators	500	
Engineering and contingencies	1,500	
Total		4,900
Grand total		8,400

TOTAL COST

Modification No. 1		$19,500
Modification No. 2		15,240
Modification No. 3		17,980
Less: fan drives furnished in #2	3,000	
Less: rebalance in #2 not required		
66 Mixing boxes @ $30	1,980	
10 Zone dampers @ $90	900	
		(5,880)
Modification No. 4		8,400
Total		55,240
Part of Modification No. 1 will be accomplished through Central Campus Energy Control Center Contract Phase I		(10,000)
Net total		$45,240

Sources of Information
on Energy Conservation

Here follows a list of sources from which further information on energy conservation may be obtained.

Societies, Associations, & Institutes

- Air-Conditioning and Refrigeration Institute, 1815 N. Ft. Myer Dr., Arlington, VA 22209
- Air Cooling Institute, P.O. Box 2121, Wichita Falls, TX 76301
- Air Diffusion Council, 435 N. Michigan Ave., Chicago, IL 60611
- Air Distribution Institute, 221 N. LaSalle St., Chicago, IL 60601
- Air Moving and Conditioning Association, 30 W. University Dr., Arlington Heights, IL 60004
- American Boiler Manufacturers Association, 1500 Wilson Blvd., Suite 317, Arlington, VA 22209
- American Consulting Engineers Council, 1155 15th St., N.W., Rm. 713, Washington, DC 20005
- American Gas Association, 1515 Wilson Blvd., Arlington, VA. 22209
- American Industrial Hygiene Association, 210 Haddon Ave., Westmont, NJ 08108
- American Institute of Architects, 1735 New York Ave., N.W., Washington, DC 20006
- American Institute of Consulting Engineers (See American Consulting Engineers Council)
- American Institute of Plant Engineers, 1021 Delta Ave., Cincinnati, OH 45208
- American National Standards Institute, Inc., 1430 Broadway, New York, NY 10018
- American Society of Heating, Refrigeration and Air-Conditioning Engineers, Inc., 345 E. 47th St., New York, NY 10017
- American Society of Mechanical Engineers, 345 E. 47th St., New York, NY 10017

- American Society of Plumbing Engineers, 16161 Ventura Blvd., Suite 105, Encino, CA 91316
- American Society for Testing and Materials, 1916 Race St., Philadelphia, PA 19103
- Associated Air Balance Council, 2146 Sunset Blvd., Los Angeles, CA 90026
- Associated General Contractors of America, 1957 E. St. N.W., Washington, DC 20006
- Better Heating Cooling Council, 35 Russo Pl., Berkeley Heights, NJ 07922
- BRAB Building Research Institute, 2101 Constitution Ave., Washington, DC 20418
- Building Owners & Managers Association International, 224 South Michigan Ave., Chicago, IL 60604
- Building Research Advisory Board, National Research Council, National Academy of Sciences-National Academy of Engineering, 2101 Constitution Ave., N.W. Washington, DC 20418
- Construction Specifications Institute, 1150 Seventeenth St., N.W., Suite 300, Washington, DC 20036
- Comevot Equipment Manufacturers, 1000 Vermont Ave., N.W., Washington, DC 20005
- Cooling Tower Institute, 3003 Yale St., Houston, TX 77018
- Edison Electric Institute, 90 Park Ave., New York, NY 10016
- Electrical Apparatus Service Association, Inc., 7710 Carondelet Ave., St. Louis, MO 63105
- Electrification Council, The, 90 Park Ave., New York, NY 10016
- Gas Appliance Manufacturers Association, Inc., 1901 N. Ft. Myer Dr., Arlington, VA 22209
- Heat Exchange Institute, 122 E. 42nd St., New York, NY 10017
- Hydronics Institute, 35 Russo Pl., Berkeley Heights, NJ 07922
- Illuminating Engineering Society, 345 E. 47th St., New York, NY 10017
- Institute of Electrical & Electronics Engineers, Inc., 345 E. 47th St., New York, NY 10017
- Instrument Society of America, Stanwix St., Pittsburgh, PA 15222
- International District Heating Association, 5940 Baum Sq., Pittsburgh, PA 15206
- Mechanical Contractors Association of America, Inc., 5530 Wisconsin Ave., Suite 750, Washington, DC 20015
- National Association of Oil Heating Service Manager, Inc., 60 E. 42nd St., New York, NY 10017
- National Association of Plumbing, Heating & Cooling Contractors, 1016 20th St., N.E., Washington, DC 20036
- National Association of Power Engineers, Inc., 176 W. Adams St., Suite 1411, Chicago, IL 60603

- National Association of Refrigerated Warehouses, 1210 Tower Bldg., 1401 K St., N.W., Washington, DC 20005
- National Coal Association, Coal Bldg., 1130 17th St., N.W., Washington, DC 20036
- National Electrical Contractors Association, 7315 Wisconsin Ave., Washington, DC 20014
- National Electrical Manufacturers Association, 155 E. 44th St., New York, NY 10017
- National Environmental Systems Contractors Association, 221 N. LaSalle St., Chicago, IL 60601
- National Insulation Contractors Association, 8630 Fenton St., Suite 506, Silver Spring, MD 20910
- National LP-Gas Association, 79 W. Monroe St., Chicago, IL 60603
- National Mineral Wool Insulation Association, Inc., 211 E. 51st St., New York, NY 10022
- National Oil Fuel Institute, Inc., 60 E. 42nd St., New York, NY 10017
- National Society of Professional Engineers, 2020 K St., N.W., Washington, DC 20006
- Producers' Council, Inc., 1717 Massachusetts Ave., Washington, DC 20036
- Refrigeration Service Engineers Society, 2720 Des Plaines Ave., Des Plaines, IL 60018
- Society of American Value Engineers (SAVE), 2550 Hargrave Dr., Smyrna, GA 30080
- Standards Engineers Society, P.O. Box 7507, Philadelphia, PA 19101
- Steam Heating Equipment Manufacturers Assoc., c/o Samuel J. Reid, Barnes & Jones, Inc., P.O. Box 207, Newtonville, MA 02160
- Thermal Insulation Manufacturers Association, Inc., 7 Kirby Plaza, Mt. Kisco, NY 10549
- Underwriters' Laboratories, Inc., 333 Pfingsten Rd., Northbrook, IL 60062
- Water Conditioning Foundation, 1780 Maple St., P.O. Box 194, Northfield, IL 60093

Local Sources

- Chapters of above-mentioned societies, associations, and institutions
- Utilities
- Chambers of Commerce
- Construction industry organizations
- Building code authorities
- Libraries
- Architectural engineers, contractors, suppliers and others with whom you work on a regular basis.

Glossary
of Terms

Ammeter — A device used to measure amperage.

Ampere — A unit of electric current equal to a flow of 1 coulomb/sec, or the steady current produced by 1 volt applied across a resistance of 1 ohm.

Aqua-Stat — A device for measuring the temperature of water within a closed container, which will automatically activate heating elements or deactivate them.

Ballast — A resistor used to stabilize the current in a given circuit.

British Thermal Unit (BTU) — A British thermal unit measures heat energy. It is defined as the quantity of heat required to raise the temperature of 1 lb. of water 1° F.

Collector — A device that collects solar radiation and converts it to heat.

Condensation — When steam or any other vapor is subjected to a change of state that reduces it to a liquid, it is said to be "condensed." Steam is condensed in a condenser or heater by extracting heat. The water formed is called "condensate."

Conduction — When heat is transmitted through a substance or from one substance to another in contact with it, without the bodies' themselves moving, the transfer is by conduction. Heat is conducted through the metal in the shell and tubes of a boiler. Substances differ widely in their ability to conduct heat. Metal is a good conductor; soot and boiler scale are very poor.

Conversion of Heat Energy and Mechanical Energy — Heat energy and mechanical energy are convertible. There is a direct relation between heat energy and mechanical energy; 778 ft-lb is equivalent to 1 BTU.

Design Heating Load — The total heat loss from a house under the most severe winter conditions likely to occur; a concept used in the design of buildings and their heating systems.

Differential Controller — A device that measures the pressure difference between two areas and can activate heating and/or cooling.

Efficiency — The efficiency of any device is the output divided by the input, sometimes stated as the useful energy divided by the energy expended. The input and output may be expressed in any energy units. They must, however, be in the same units.

Energy — Energy is the ability to do work. Mechanical energy is expressed in foot-pounds or horsepower hours; electrical energy, in kilowatt-hours; and heat energy, in British Thermal Units.

Evaporation — The process of changing a liquid into a vapor or a gas is known as "vaporization." This usually is accomplished by the application of heat.

Factor of Evaporation — If 970.3 BTU are added to 1 lb. of water that is at atmospheric pressure and 212° F, it will be converted into steam, and the steam will be at atmospheric pressure and 212° F. This is termed "evaporation from and at 212° F," or briefly "from and at." The heat added to a pound of water by the boiler (from the time it enters until it leaves as steam) is divided by the 970.3; the quotient is the factor of evaporation.

Ferrous — A compound or metal which contains iron.

Force — Force is what produces, or tends to produce, motion. The force on the piston of a steam engine produces motion. The force exerted on the head of a steam engine produces motion. The force exerted on the head of a steam boiler does not produce motion, but it tends to; both are examples of force.

Geothermal Energy — Energy which is generated by harnessing heat from the earth's interior.

Heat Exchanger — A device, such as a coiled copper tube immersed in a tank of water, that is used to transfer heat from one fluid to another through a separating wall.

Heat Storage — A device or medium that absorbs collected solar heat and stores it for use during periods of inclement or cold weather. Solar systems could store heat in a tank containing water, rocks, or eutectic salts.

Horsepower — A unit of power (hp) that tells the rate at which work is being performed, namely, 33,000 ft-lb per min.

Horsepower-Hour — A horsepower-hour is 1 hp of energy expended continuously for 1 hr.

Kilowatt — A unit of electric energy equal to 1,000 watts. For direct current, watts = amperes × volts; for alternating current, watts = amperes × volts × power factor.

Kilowatt-Hour — One kw of energy expended continuously for one hour.

Law of Conservation of Energy — The amount of energy in existence is constant. The machines that we build and operate do not produce energy; they merely change it from one form to another.

Photovoltaic Cells — Semiconductor (solid state) devices that convert solar energy directly into electricity with no moving parts.

Polymers — Chemical compounds or mixtures in which two or more small molecules combine to form larger molecules which contain repeating structural units of the original small molecules.

Power — The rate at which work is done. Foot-pounds express work, but the rate or time required determines the power: 33,000 ft-lb per min is a horsepower.

Pulverized Fuels — Fossilized fuels, *e.g.,* coal, which have been crushed into small particles. This type of fuel is usually associated with suspension burning.

Radiation — The transmission of heat without the use of a material carrier. The earth receives heat from the sun, and most of that distance the heat travels through a vacuum. When a furnace door is open, the heat can be felt even though air is being pulled into the furnace through the door. The lower rows of boiler tubes receive much heat by radiation. Radiated heat is very similar to visible light; both travel at the same speed, 186,000 miles per second.

Retrofitting — The addition of equipment and/or system to an existing building.

Solar Rights A legal issue concerning the right of access to sunlight, also known as Sun Rights of Solar Access.

Specific Heat — Different substances have different heat capacities. In fact, the heat capacity of some substances changes as the temperature changes. By the definition of a BTU, the heat content of a pound of water per degree Fahrenheit is 1. The specific heat of any substance is the heat required to raise 1 lb of it 1° F.

Temperature — Temperature of a substance must be distinguished carefully from the heat content. Temperature is thermal pressure and is a measure of the ability of a substance to give or recieve heat from another.

Thermal Efficiency — This refers to heat engines and is the output expressed in BTU divided by the input expressed in BTU.

Thermosphon — A solar heater using no pumps. As water is heated it moves through a panel to storage, then from storage through the panel.

Voltmeter — A device used to measure voltage.

Index